Engineering Materials

More information about this series at http://www.springer.com/series/4288

Emilio I. Alarcon · May Griffith
Klas I. Udekwu
Editors

Silver Nanoparticle Applications

In the Fabrication and Design of Medical and Biosensing Devices

Springer

Editors
Emilio I. Alarcon
Bio-nanomaterials Chemistry
and Engineering Laboratory,
Cardiac Surgery Research
University of Ottawa Heart Institute
Ottawa
Canada

Klas I. Udekwu
Swedish Medical Nanoscience Center
Karolinska Institutet
Stockholm
Sweden

May Griffith
Integrative Regenerative Medicine Centre
Linköping University
Linköping
Sweden

ISSN 1612-1317 ISSN 1868-1212 (electronic)
Engineering Materials
ISBN 978-3-319-11261-9 ISBN 978-3-319-11262-6 (eBook)
DOI 10.1007/978-3-319-11262-6

Library of Congress Control Number: 2015930508

Springer Cham Heidelberg New York Dordrecht London

Printed on acid-free paper

Springer International Publishing AG Switzerland is part of Springer Science+Business Media
(www.springer.com)

To Alonso for not giving up and Madleen for her love; and in memory of Alexander Y.N.

To Malcolm, Meagan, Marisa, Pip and Button; and in memory of little Rowley

To Ruth, Sofia, Ben, Lena and chi'm

Preface

Nanomaterials bear the promise of revolutionizing the development of biomaterials for the medical sciences and biosensing. However, prior to safe and efficacious translational applications of such materials in the clinic, comprehension of the nature of nanoparticles and the properties they impart to the materials that they are incorporated into them, is necessary. Hence, multidisciplinary collaboration amongst biologists, chemists, engineers, physicists, and clinicians is critical for designing the next generation of nanomaterials with improved biological activity and regenerative properties, and for moving these along the translational pipeline from "bench to bedside."

Silver nanoparticles, in particular, have a special, almost unique, place among nano-sized materials. This is due to their unique and multi-functional properties that include their archetypical antimicrobial activity, excellent thermoplasmonic capabilities, and superior surface Raman properties. This book, authored by active researchers, reviews the latest research on silver nanoparticles and nanomaterials around the globe. We provide an overview of the current knowledge on the synthesis, uses, and applications of nanoparticulate silver. In short, students and researchers in the field will gain an up-to-date understanding of what silver nanoparticles are, their current uses, and future challenges and horizons of these nanomaterials in the development of new materials with improved properties.

Emilio I. Alarcon
May Griffith
Klas I. Udekwu

Acknowledgments

The editors would like to express their gratitude to the authors of this book; without their valuable contribution this endeavor would not have been possible. Also, the editors would like to express their thankfulness to Dr. Rashmi Tiwari-Pandey at the Division of Cardiac Surgery—Biomaterials and Regeneration Program, University of Ottawa Heart Institute, for her help during the final stages of formatting and proofreading of the book.

Emilio I. Alarcon
May Griffith
Klas I. Udekwu

Contents

Contributors

Emilio I. Alarcon Bio-nanomaterials Chemistry and Engineering Laboratory, Division of Cardiac Surgery, University of Ottawa Heart Institute, Ottawa, Canada; Centre for Catalysis Research and Innovation, University of Ottawa, Ottawa, Canada

Hasitha de Alwis Weerasekera Department of Chemistry and Centre for Catalysis Research and Innovation, University of Ottawa, Ottawa, Canada

Hanan Anis School of Electrical Engineering and Computer Science, University of Ottawa, Ottawa, ON, Canada

Claudio D. Borsarelli Laboratorio de Cinética y Fotoquímica (LACIFO), Centro de Investigaciones y Transferencia de Santiago del Estero (CITSE-CONICET), Universidad Nacional de Santiago del Estero (UNSE), Santiago del Estero, Argentina

Spyridon Gkotzis Swedish Medical Nanoscience Centre, Department of Neuroscience, Karolinska Institutet, Stockholm, Sweden

María González-Béjar Instituto de Ciencia Molecular (ICMol)/Departamento de Química Orgánica, Universidad de Valencia, Valencia, Paterna, Spain

May Griffith Integrative Regenerative Medicine Centre, Department of Clinical and Experimental Medicine, Linköping University, Linköping, Sweden; Swedish Medical Nanoscience Centre, Department of Neuroscience, Karolinksa Institutet, Stockholm, Sweden

Altaf Khetani School of Electrical Engineering and Computer Science, University of Ottawa, Ottawa, ON, Canada

Thien-Fah Mah Department of Biochemistry, Microbiology and Immunology, Faculty of Medicine, University of Ottawa, Ottawa, Canada

Ali Momenpour School of Electrical Engineering and Computer Science, University of Ottawa, Ottawa, ON, Canada

Natalia L. Pacioni INFIQC, CONICET and Departamento de Química Orgánica-Facultad de Ciencias Químicas-Universidad Nacional de Córdoba, Ciudad Universitaria, Córdoba, Argentina

Valentina Rey Laboratorio de Cinética y Fotoquímica (LACIFO), Centro de Investigaciones y Transferencia de Santiago del Estero (CITSE-CONICET), Universidad Nacional de Santiago del Estero (UNSE), Santiago del Estero, Argentina

Kevin Stamplecoskie Radiation Laboratory, University of Notre Dame, South Bend, IN, USA

Vidhu S. Tiwari SRM University, Sonepat–Kundli Urban Complex Sonepat, Haryana, India

Klas I. Udekwu Swedish Medical Nanoscience Centre, Department of Neuroscience, Karolinksa Institutet, Stockholm, Sweden

Alicia V. Veglia INFIQC, CONICET and Departamento de Química Orgánica-Facultad de Ciencias Químicas-Universidad Nacional de Córdoba, Ciudad Universitaria, Córdoba, Argentina

Silver Nanoparticles: From Bulk Material to Colloidal Nanoparticles

Kevin Stamplecoskie

Abstract Metals exhibit interesting optical properties, especially in comparison to molecules and semiconductors. In contrast to molecules and semiconductors, metals support plasmons, which are a collective oscillation of many electrons in the material. When the size of these metal nanoparticles is small (<100 nm), these plasmon absorbances occur in the visible region of the electromagnetic spectrum, giving rise to colored solutions. One of the unique characteristics of plasmon excitation is the conversion of light energy into extreme and highly localized heating at the surface of these particles. Excitation of plasmons by both pulsed (i.e. lasers) and continuous (i.e. sunlight) excitation and the effects of plasmon excitation on the surrounding material are discussed in this chapter. The potential for using these materials in photothermal therapy for ailments such as cancer is also discussed in terms of the unique properties of these metals, related to plasmon excitation.

Keywords Nanomaterials · Silver nanoparticles · Plasmon

1 Introduction

The existence of metal nanoparticles is not new; they have been around since ancient times. The most famous example is the Lycurgus Cup, made in the 4th century AD. The glass used is colored with gold nanoparticles and appears red when lit from behind (light through it) and green when lit from the front [1]. The fact that gold nanoparticles and their plasmon absorption were responsible for the pretty colors in this stained glass was certainly not understood, but nevertheless nanoparticle synthesis has a very ancient and rich history. Similarly, while the past few decades have experienced the resurgence in the use of silver nanoparticles

K. Stamplecoskie (✉)
Radiation Laboratory, University of Notre Dame, South Bend, IN, USA
e-mail: kstamp4069@gmail.com

E.I. Alarcon et al. (eds.), *Silver Nanoparticle Applications*, Engineering Materials,
DOI 10.1007/978-3-319-11262-6_1

(AgNP) in biomedical applications, silver also has an ancient history for medicinal purposes. The medicinal effects of silver date back to when ancient Romans and Phoenicians stored drinking water in containers made of silver [2]. Silver has, throughout history been continually used in medicine for its antibacterial properties. It now finds applications embedded in clothing and fabrics, surgical grade steel, deodorants, toothpaste, toys, humidifiers and much more; used to slow the growth of unwanted bacteria [3]. Interestingly, for so many years silver has been used without a clear understanding of the mechanism of antibacterial activity. Only recently has there been further understanding into how AgNP act as antibacterial agents (see Chap. "Anti-microbiological and Anti-infective Activities of Silver").

Metal nanoparticles such as gold, silver and copper also display unique and interesting optical and electronic properties. These physical properties are the main topic of this chapter. Everyone is familiar with the color of metals such as gold, silver and copper, but the colors of these metals are very different when the particle diameter becomes very small (<100 nm). Solutions of nanoparticles, particles embedded in transparent matrices (i.e. glass) or nanoparticles supported on other solids absorb visible wavelengths of light, giving rise to colors that can span the visible spectrum. These colors and absorption properties of metals are due to plasmon excitation.

The interesting optical properties of metal nanostructures have driven a surge of research interest over the past couple of decades for applications in molecular sensing [4], creating ultrafast optoelectronics [5] and biomedicine for targeting and killing cancer cells [6]. The remainder of this chapter will focus on how some common metal nanostructures can be excited by light, with an emphasis on AgNP. We will discuss the unique relaxation processes that follow plasmon excitation as well as ways in which the optical properties of metals have been exploited in biomedical application.

2 Excitation of Metal Nanoparticles

Undergraduate courses have educated chemists and physicists, developing an in depth understanding of the optical excitation and relaxation of most materials. The electronic and vibrational motions of molecules are treated separately and studied spectroscopy with characteristics such as vibrational and electronic energy levels. Excitation of molecules and semiconductors with electromagnetic radiation is due to excitation of electrons to higher energy levels. There is a defined timescale for the average lifetime of the excited state that is governed by the combination of radiative and non-radiative processes e.g. intersystem crossing, internal conversion, fluorescence, etc. [7]. This theory breaks down when discussing materials like metals that have a relatively high number of electrons close in energy to a large number of available empty states, where electrons can freely transfer between states at room temperature [8]. In other words, materials with many available electronic states directly above the Fermi level, display properties such as the high conductivity seen for metals.

Mie theory was developed to explain the unique optical properties of light scattering and absorption displayed by metals. It provides an understanding that explains the optical properties of metal nanoparticles in a fundamentally different way from conventional molecular photophysics [9]. According to Mie theory, the choice of metal, as well as size, shape, surrounding matrix, surface bound molecules and degree of aggregation of the particles determines the energy range (frequency) of light that can excite plasmons. For example, the major (dipole) absorption of spherical nanoparticles is predicted by Mie theory to be approximately 400 nm. Lager AgNP and those with different shapes, however, absorb different wavelengths of light due to other absorption modes [10]. Colloidal solutions of spherical copper and gold nanoparticles, however, are orange and red because their plasmon absorption maxima occur at approximately and 530 nm and 580 nm, respectively.

So what is a plasmon or plasmon absorption, really? To answer this question we will begin by discussing the properties of bulk metals, that also support plasmons. A large, flat piece of metal can be viewed as an infinite, periodically arranged positive charges (nuclei) with loosely bound cloud of electrons held by a coulombic attraction. When light of an appropriate frequency interacts with the surface of a metal, the electric field component of light couples with the electrons in the metal causing an instantaneous displacement of the electron density. The light is absorbed forming a periodic fluctuation of positive and negative charges called 'surface plasmon polaritons', as illustrated in Fig. 1. The nuclei serve as a restoring force on the electrons, where the magnitude of this restoring force is a function of the exact nuclei (the type of metal used). There is a strong local electromagnetic field produced by these rapid, and coherent oscillating electrons that extends into the metal and surrounding medium. For this reason, the frequency that can be used to excite plasmon absorptions is a function of both the metal and the dielectric medium surrounding it. On a bulk metal, these surface plasmons propagate along the surface for a particular distance. For typical metals used as waveguides and sensors, these surface plasmon excitations occur in the infrared region of the electromagnetic spectrum. The rest of the light is reflected (not absorbed into plasmon excitations). For this reason, metals are reflective, which is why silver is used for mirrors, and why pieces of these metals appear shiny.

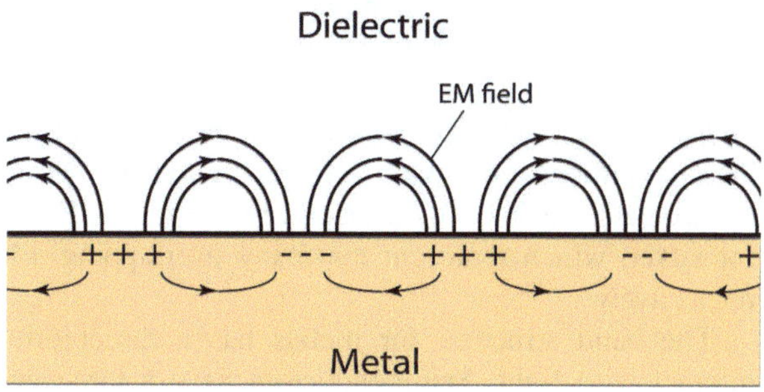

Fig. 1 Schematic illustration of propogating surface plasmon polaritons in a bulk metal as well as the resultant electromagnetic field in both the dielectric (surroundings) and metal

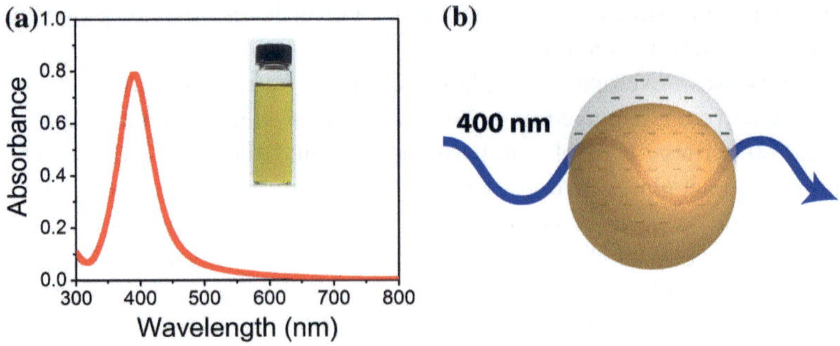

Fig. 2 **a** Absorption spectrum of an aqueous solution of spherical AgNP as well as an image of this colloidal solution (inset). **b** Schematic illustration of plasmon excitation causing an instantaneous collective oscillation of electrons

So how is this relevant to silver nanoparticles? In the bulk metal example, the surface of the metal is a flat plain and the size of the metal was also infinite with respect to the light wave. Metal nanoparticles, however, can be viewed as small metal slabs that are now smaller than the wavelength of light used to excite them. This has two dramatic differences with bulk metals; (1) all of the loosely bound electrons in nanoparticles are instantaneously excited by light, and (2) the surface curvature of a nanoparticle is no longer flat with respect to the wavelength of light. For this reason, the plasmon absorptions of nanoparticles occur in the visible region of the electromagnetic spectrum, giving rise to the multitude of colors displayed by metal nanoparticles. Also, the plasmons of nanoparticles cannot propagate since they are confined to the particle. For this reason, nanoparticle plasmons are commonly referred to as 'localized surface plasmons' (LSP), depicted in Fig. 2b.

AgNP (like other metals) can support many different plasmon modes, especially for different shapes of nanoparticles, where the isotropy of the particle is broken. Fig. 2a shows a typical absorption spectrum for spherical ~3.3 nm colloidal AgNP as well as an image of a solution of these particles, showing the yellow color due to the ~400 nm dipolar plasmon absorption. The absorption spectrum can be tuned by simply controlling the shape of the crystals, giving absorption maxima that span the full visible spectrum and colloidal solutions of many different colours. For different shapes and sizes, the plasmons absorptions are still localized to the particle, however, the higher order modes involved for these different particles have more complex electron distributions than the simple oscillating linear dipole mode for spheres [10].

Density of states diagrams are commonly used by materials scientists to describe the distribution of electrons and electronic states of materials. The overlap of many electronic states is called a 'band'. For example, Fig. 3 shows the 4d and 5sp bands for silver, which represent the many overlapping 4d and hybridized 5sp orbitals, respectively.

The band structure for metals has a direct effect on the way these metals interact with light. Semiconductors have a large energy gap between electronic

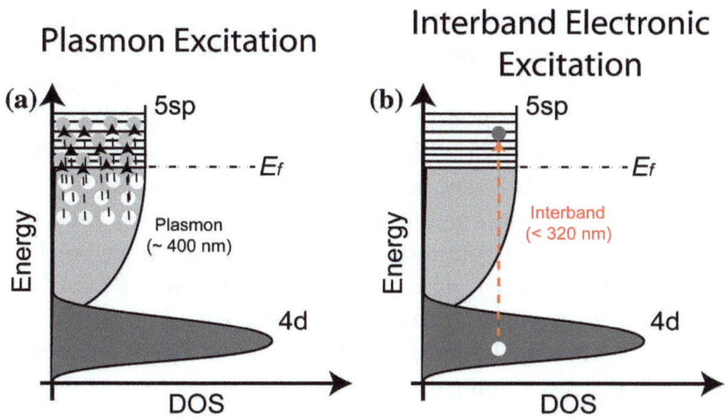

Fig. 3 Schematic illustration of the density of states of silver under plasmon excitation where a collection of electrons are excited (**a**) and electronic excitation where a single electron is excited (**b**)

states that are occupied by electrons and ones that are not. The energy separation between the filled 'valence' band and the empty 'conduction' band in semiconductors is called the 'band gap'. Metals, like silver, however, have no band gap. Filled electronic states of the 5sp band overlap in energy with unfilled states of the same band. Therefore, the instantaneous displacement of electrons in plasmon absorption is physical (as depicted in Fig. 2b) but can also be described as a collective excitation of many electrons to slightly higher energy, as depicted in the density of states scheme in Fig. 3a. In addition to plasmon excitation, AgNP it is also possible to have electronic excitation as depicted by the 'interband transitions' in Fig. 3c, where individual electrons can be excited to higher energy levels. While this electronic excitation overlaps with plasmon excitation for metals such as gold and copper, electronic excitations for silver occur at higher energies than plasmon excitation (less than 320 nm), as illustrated in Fig. 3b [11].

From Mie theory we find that the ability of a material to support plasmon absorptions (polarizability) is given by Eq. 1;

$$\alpha = 3\varepsilon_0 V \frac{\varepsilon - \varepsilon_m}{\varepsilon - 2\varepsilon_m} \tag{1}$$

where, α is the polarizability, ε_0 is the permittivity of a vacuum, V is the volume of the particle, ε is the permittivity of the material, and ε_m is the permittivity of the medium surrounding the particle. From the numerator in Eq. 1, a maximum polarizability occurs when the permittivity of the material is approximately -2 times the permittivity of the medium. Since the permittivity of most materials and gases is positive and approximately unity, this means that the permittivity of the material that supports a plasmon must be negative. In short, metals have a negative permittivity, and this is the reason we see plasmon absorption for metals.

Also important to note is that metals are not all created equal. One of the particular advantages of using AgNP over other metals like gold and copper is that AgNP have a higher cross section for absorption than the others. This has to do with the relative permittivity of the metals where silver has a more negative real part of the permittivity than Au, for instance. As an example, the extinction coefficient for plasmon absorption for a 20 nm AgNP has been experimentally determined to be $4.75 \times 10^9\ M^{-1}\ cm^{-1}$, whereas for AuNP the value is $1 \times 10^9\ M^{-1}\ cm^{-1}$ [12]. The absorption cross-section for Ag plasmons is almost five times that of Au. For this reason AgNP are often chosen over other metals in applications when plasmon excitation is important, such as surface enhanced Raman spectroscopy and photothermal therapy. To further highlight the absorption properties of AgNP, dye extinction coefficients are on the order of $\sim 10^{4-5}\ M^{-1}\ cm^{-1}$, whereas the plasmon absorption of AgNP is several orders of magnitude higher, depending on particle size.

The oscillating displacement of electrons with respect to positively charged nuclei that occurs upon plasmon excitation, has the effect of generating a strong electric field very close to the surface of the particles. The implications of the strong oscillating dipole are further discussed with respect to enhancing biochemical sensing and photochemical reactions in Chap. "Surface Enhanced Raman Scattering (SERS) Using Nanoparticles".

Furthermore, metals have many electrons as compared to molecules, which means that the cross-section for exciting metal nanoparticles with either linear or multi-photon excitation is very high [8]. This is important for many applications where intense light can damage the biological systems, since lower light intensities can be used to excite metal nanoparticles without directly damaging the cells or biological material of interest. In addition to single photon absorption, plasmons have extremely high cross-section for multi-photon absorbance. This non-linear process occurs when multiple low energy photons are simultaneously absorbed. For example, you can excite AgNP with 400 nm light, but they can also be excited by simultaneous absorption of two 800 nm photons. The advantages of being able to excite metal nanoparticles with near infrared light for imaging, sensing and photodynamic therapy are discussed further at the end of this chapter.

3 Relaxation Processes Following Nanoparticle Excitation

In the previous section, the excitation of metal nanoparticles with EM radiation (light) was discussed. Understanding the effects of excitation on the nanoparticle and its surroundings is of utmost importance, especially when incorporating nanomaterials into biologically relevant systems. The relaxation dynamics in excited metals are extremely fast in comparison to molecules. This is also a result of the small energy separation between excited electrons and the ground state. Kasha's rule is an important foundation in molecular photochemistry, and it states that the rate of an electronic relaxation is inversely proportional to the difference in energy between energy states [7]. This means that, for metals with overlapping filled and unfilled orbitals at the Fermi level energy (E_F in Fig. 3), the rate of relaxation must

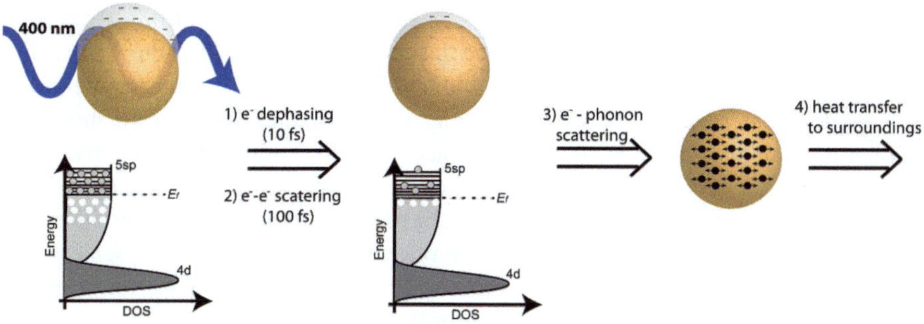

Fig. 4 Illustration of the processes involved in relaxation after plasmon excited of metal nanoparticles including (*1*) dephasing of coherent electron oscillation, (*2*) e^-—e^- scattering, (*3*) e^-—phonon scattering, (*4*) heat transfer to the medium

be very fast. Only with the recent advances in ultrafast lasers has it been possible to experimentally probe the excited state lifetime and relaxation dynamics for metals. What follows is a description of the processes involved in relaxation of plasmons in chronological order following excitation. The processes are summarized in the illustrations in Fig. 4.

Plasmon excitation causes many electrons to rapidly oscillate, in phase with each other, at the same frequency as the light used to excite them. These oscillating electrons (with respect to positively charged atomic nuclei) generate a strong oscillating electric dipole. During this time, electronic and vibrational transitions are enhanced for molecules within the electric field of the excited particles. The enhanced field is responsible for the well-known surface enhanced Raman scattering (SERS) and surface enhanced infrared absorption spectroscopy (SEIRS) effects. Any electronic or vibrational transition involving a transition dipole can be enhanced in the induce electric field of an excited plasmon. While SERS and SEIRS have become commonly observed effects, plasmon enhanced electronic transitions like absorption and fluorescence also occur [13].

The coherent oscillating electrons collide with one another causing the electrons to rapidly go out of phase with one another. This 'electron dephasing' (1) causes a non-Fermi distribution of electrons and it occurs within the ~10 fs of excitation. The excited electrons further scatter with each other eventually leading to a more randomized hot electron distribution. The 'electron-electron' scattering (2) occurs within the first ~100 fs after excitation. Most spectroscopic techniques use lasers with greater than 100 fs pulse widths, so these first electron relaxation steps are very rarely observed experimentally.

Hot electrons collide with and transfer vibrational energy (in the form of phonons) to nuclei in a so-called 'electron-phonon' decay (3). The timescale for this electron-phonon decay depends strongly on the excitation wavelength and intensity. A nanoparticle under higher intensity excitation (photon density) has to dissipate more energy from hot electrons into phonons, and so this process is longer-lived for more excited particles but generally occurs on the timescale of ~1 ps. The lifetime of electron-phonon decays can be easily determined using ultrafast transient absorption techniques with femtosecond lasers of ~100 fs pulse durations. This dependence of the hot electron lifetime with excitation intensity is one property that is very different from molecules. Changing the excitation intensity in molecular studies is analogous to controlling the concentration of molecules that are in the excited state, but every excited molecule behaves in the same way as it decays and so the lifetime of this process is independent of excitation power. Plasmon excitation, on the other hand, at varied excitation powers, is like an energy scale where more or less energy is put into each particle. For this reason, dissipation of energy into heat takes longer for more excited particles.

Temperature is a measure of the energy in that system in the form of vibration of nuclei. Put simply, electron-phonon decay is the transfer of electronic energy to vibrational motion. Therefore, excited metal nanoparticles become very hot within ~1 ps of excitation. In fact, metals display no observable emission, with ~100 % of the absorbed energy converted to heat. The high cross-section for plasmon excitation was already discussed. This strong ability to absorb light combined with the almost perfect conversion of this light to heat makes nanoparticles a great source for delivering a lot of localized heating.

The final process in plasmon relaxation is the transfer of heat to the surroundings (4). The heat transfer depends strongly on excitation intensity (how hot the nanoparticle is) as well as the thermal conductivity of the medium. Heat transfer to the surroundings occurs in hundreds of picoseconds to nanoseconds, following excitation. The ability to absorb a lot of light, and release that energy locally in a short amount of time (<1 ns), is a unique property of metal nanoparticles.

When investigating the effects of plasmon excitation, particles are often under continuous (not pulsed) excitation. For example, nanoparticles incorporated in a topical cream for a patient are under continuous excitation from sunlight or lights in a room. In laboratory studies, imaging with nanoparticles incorporated into tissues or cells is often done under continuous lamp irradiation or with continuous wave lasers. The above discussion of the timescales of the different relaxation processes is predicated on the use of pulsed excitation. That is, these processes have finite lifetimes when one can use a short excitation pulse, and monitor the nanoparticle relaxation with ultrafast spectroscopy. When the nanoparticles are under continuous excitation, however, the notion of 'lifetimes' for these processes is lost. It is important to understand that, when using continuous excitation, nanoparticles are constantly being excited and continually going through various stages of the relaxation process. This means that there is a continuous electromagnetic field generated around nanoparticles that can enhance optical and vibrational transitions in neighbouring molecules. Furthermore, these nanoparticles under constant irradiation are continually converting that absorbed light to heat. The nanoparticles effectively act as a point source of intense and continuous heating to their surroundings. Figure 5 illustrates the differences between continuous wave (CW) and laser pulsed excitation with respect to the temporal and spatial profiles of heating near an excited nanoparticle. There are many factors such as light intensity, pulse duration, choice and size/shape of the nanoparticle, that all

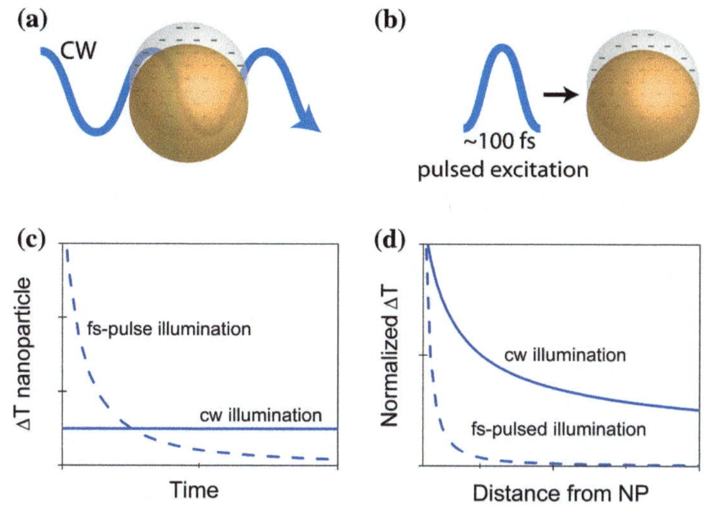

Fig. 5 Schematic illustration of plasmon excitation by either CW illumination **a** or femtosecond laser pulse excitation. **b** The temperature of a nanoparticle versus time **c** as well as the temperature of the surrounding medium with distance from the nanoparticle surface normalized to the maximum temperature at the surface **d** when excited with equivalent power of either laser pulsed excitation or CW illumination

affect the exact temperatures in Fig. 5c, d. For simplicity, Fig. 5 simply illustrates the general trend in differences between CW and pulsed excitation. For further reading and to mathematically solve exact temperatures as a function of time and distance, readers are referred to the work of Baffou and Quidant [14] and Baffou and Rigneault [15].

In the illustration in Fig. 5, we assume that the nanoparticle is under the same overall power, e.g. 1 mW/cm^2, whether that power is compressed into a 100 fs pulse or spread out over the entire second. The near perfect quantum efficiency of conversion of absorbed light to heat was discussed above. With this in mind, and since energy is always conserved, the area under each of the curves in either Fig. 5c or d is the same for both illumination conditions. Figure 5d has been normalized to the maximum temperature at the surface only to allow both decay curves to be observable on the same scale. It is not surprising that, when excited with an intense pulse of light, nanoparticles initially experience a high temperature that dissipates, whereas nanoparticles excited with the same power evenly spread out in time equilibrate at a constant, elevated temperature (Fig. 5c). What may be more surprising is that, when excited with a short laser pulse, the medium near the nanoparticle surface experiences and extreme heating that is not transferred to long distances in the medium when compared with continuous irradiation (Fig. 5d). Overall, pulsed excitation can be viewed as a method for delivering heat with high temporal and spatial control.

Under both CW or laser pulsed excitation conditions, it is perfectly valid to think of nanoparticles as point sources for enhanced EM fields and extreme heating. In the following section, some examples of utilizing the extreme heating and unique properties of metals, for photothermal and photodynamic therapy, are discussed.

4 Photothermal Therapy with 'Hot' Nanoparticles

Photothermal therapy using metal nanoparticles exploits the extreme light to heat conversion of metal nanoparticles for selective destruction of tumours. Over the past several decades there has been significant progress in the synthesis of nanoscale materials with excellent control over size and composition and surface functionality. This progress has lead to a deeper understanding of the size and composition dependent optical properties of these materials [16]. Furthermore, the advances in synthesizing materials with tailored surface functionality has been pivotal to incorporating nanoparticles into tissues, cells and biomimetics for various medicinal purposes.

The advancement of lasers of the past few decades has lead to a better understanding of the excited state dynamics of excited plasmons, but it has also lead to the development of many other fields of research. Photothermal therapy for cancer treatment is one such area. Conventional photothermal therapy uses dye molecules. These dyes can be selectively excited by lasers and release the absorbed energy as heat, which subsequently damages and destroys tumours. Most photothermal therapies using lasers rely on invasive endoscopes and catheters that deliver fiber optic cables to tumour cells. An alternative approach is to use immunotargeting, where the absorbing species is made to concentrate due to uptake selective uptake through molecular recognition. In this second approach, large areas of tissues can be irradiated with light that causes them no particular damage, except where the target molecules are located on tumour cells. This approach uses absorbing species that have specificity for tumour cells.

Controlling the surface functionalization of nanoparticles, they too can be tailored to accumulate in tumor cells. Furthermore, the optical properties of silver nanoparticles, with their high absorption cross-section and almost perfect conversion of absorbed light into heat, make them a prime candidate for photothermal therapy. This new strategy of using plasmonic nanostructures for delivering highly localized heating and destruction of tumour cells is called plasmonic photothermal therapy (PPTT) [17].

Nanoparticles can be functionalized with tumor-targeting molecules and selectively accumulate in tumors. For example, some malignant tumors like breast cancer tumors, over-express epithelial growth factor receptors (EGFR). Metal nanoparticles can be functionalized with anti-EGFR antibodies, and when mixed with malignant cells, the antibody-receptor interactions, concentrates nanoparticles on tumor cells. Subsequent plasmon excitation from laser irradiation has been shown to be very effective in selectively killing malignant tumors in this way. Figure 6 shows a schematic illustration of PPTT [17].

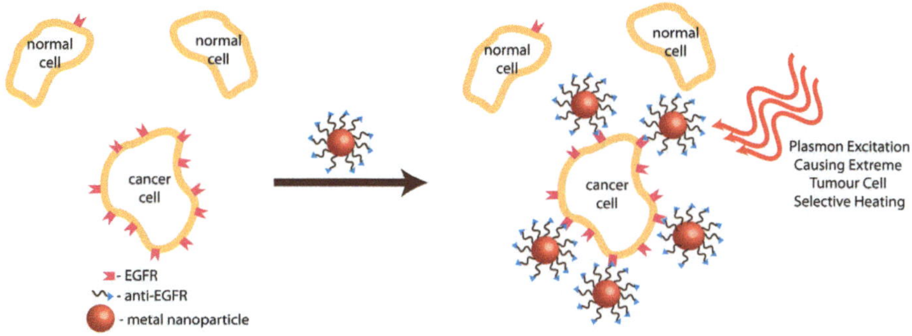

Fig. 6 PPTT scheme including selective concentration of nanoparticles functionalized with antibodies onto tumour cells through antibody-receptor interaction followed by laser plasmon excitation for extreme heating causing cell death

Most photothermal therapy strategies currently use gold nanoparticles due to their superior stability under many different conditions, in biological relevant mediums. Biomolecule conjugated AgNP synthesis, with good stability, is rapidly developing and opening the door to using AgNP in more PPTT strategies [3, 18]. The superior optical properties of AgNP in comparison to gold make it even more attractive for higher efficiencies of light to heat conversion in PPTT.

Most people have either put a flashlight in their mouth or behind their hand. What you see is a red glow of light that passes through the tissue. This is because red light has a longer penetrating depth than blue light, so that some of the red light makes it through your skin or mouth tissue. In photothermal therapy, the wavelength of excitation is an important factor then, since only red light can penetrate deep into tissue to excite the absorbing species (that subsequently releases heat). It was briefly mentioned in the previous section, that the multi-photon cross-section for metal nanoparticles is very high in comparison to dyes. This becomes even more relevant in PPTT treatments when it is desirable to use near infrared laser excitation. Metal nanoparticles are particularly effective at absorbing NIR light, and converting that light to heat.

5 Concluding Remarks

The optical properties of AgNP are very different from molecules that are more commonly understood. The different interaction that light has with AgNP opens many opportunities for using these particles. Molecular sensing strategies using the enhanced electromagnetic fields around particles, fast optoelectronics that use the ultrafast plasmon collective oscillations (~10 fs lifetime) and applications like chemical catalysis and photothermal cancer therapy that exploit the extreme and localized heating of AgNP, are only a few examples of how the unique optical properties of AgNP can be used.

References

1. Goesmann, H., Feldmann, C.: Nanoparticulate functional materials. Angew. Chem. Int. Ed. **49**, 2–36 (2010)
2. Alexander, J.W.: History of the medical use of silver. Surg. Infec. **10**(3), 289–292 (2009)
3. Alarcon, et al.: The biocompatibility and antibacterial properties of collagen-stabilized, photochemically prepared silver nanoparticles. Biomaterials **33**(19), 4947–4956 (2012)
4. Rosi, N.L., Mirkin, C.A.: Nanostructures in Biodiagnostics. Chem. Rev. **105**(4), 1547–1562 (2005)
5. Stockman, M.I.: The spaser as a nanoscale quantum generator and ultrafast amplifier. J. Opt. **12**, 024004 (2010)
6. Lewinski, N., Colvin, V., Drezek, R.: Cytotoxicity of nanoparticles. Small **4**(1), 26–49 (2008)
7. Turro, N.J., Ramamurthy,V., Scaiano, J.C.: Modern molecular photochemistry of organic molecules, University Science Books, 1100 (2009)
8. Hartland, G.V.: Optical studies of dynamics in noble metal nanostructures. Chem. Rev. **111**(6), 3858–3887 (2011)
9. Kelly, K.L., et al.: The optical properties of metal nanoparticles: the influence of size, shape, and dielectric environment. J. Phys. Chem. B **107**, 668–677 (2003)
10. Stamplecoskie, K.S., Scaiano, J.C.: Light emitting diode irradiation can control the morphology and optical properties of silver nanoparticles. J. Am. Chem. Soc. **132**(6), 1825–1827 (2010)
11. Henglein, A.: Physichochemical properties of small metal particles in solution: "Microelectrode" reactions, chemisorption, composite metal particles, and the atom-to-metal transition. J. Phys. Chem. US **97**(21), 5457–5471 (1993)
12. Link, S., Wang, Z.L., El-Sayed, M.A.: Alloy formation of gold-silver nanoparticles and the dependence of the plasmon absorption on their composition. J. Phys. Chem. B **103**(18), 3529–3533 (1999)
13. Anger, P., Bharadwaj, P., Novotny, L.: Enhancement and quenching of single-molecule fluorescence. Phys. Rev. Lett. **96**(11), 113002 (2006)
14. Baffou, G., Quidant, R.: Nanoplasmonics for chemistry. Chem. Soc. Rev. **43**(11), 3898 (2014)
15. Baffou, G., Rigneault, H.: Femtosecond-pulsed optical heating of gold nanoparticles. Phys. Rev. B **84**(3), 13 (2011)
16. Mohamed, M.B., et al.: Hot electron and phonon dynamics of gold nanoparticles embedded in a gel matrix. Chem. Phys. Lett. **343**(1–2), 55–63 (2001)
17. El-Sayed, I., Huang, X., El-Sayed, M.: Selective laser photo-thermal therapy of epithelial carcinoma using anti-EGFR antibody conjugated gold nanoparticles. Cancer Lett. **239**(1), 129–135 (2006)
18. Alarcon, et al.: Human serum albumin as protecting agent of silver nanoparticles: role of the protein conformation and amine groups in the nanoparticle stabilization. J. Nanopart. Res **15** 1–14 (2013)

Synthetic Routes for the Preparation of Silver Nanoparticles

A Mechanistic Perspective

Natalia L. Pacioni, Claudio D. Borsarelli, Valentina Rey and Alicia V. Veglia

Abstract In this chapter, we revise some of the most relevant and widely used synthetic routes available for the preparation of metallic silver nanoparticles. Particular emphasis has been focused in the rationale involved in the formation of the nanostructures, from the early metallic silver atoms formation, passing by atoms nucleation and concluding in the growth of silver nanostructures. We hope the reader will find in this chapter a valuable tool to better understand the relevance of the experimental conditions in the resulting silver nanoparticle size, shape and overall properties.

1 Introduction

Silver nanoparticles (AgNP) are already part of our daily life, being present in clothes (e.g. in socks); household and personal care products, mainly due to their antimicrobial properties [1, 2], see Chaps. "Biomedical Uses of Silver Nanoparticles: From Roman Wine Cups to Biomedical Devices" and "Anti-microbiological and Antiinfective Activities of Silver".

Furthermore, as discussed in the previous chapter, their unique physical and electronic properties make them excellent candidates for different applications e.g. Surface Enhanced Raman Spectroscopy (SERS) [3–9]. The optical properties of AgNP depend

N.L. Pacioni (✉) · A.V. Veglia
INFIQC, CONICET and Departamento de Química Orgánica-Facultad de Ciencias Químicas-Universidad Nacional de Córdoba, Ciudad Universitaria, Edificio Ciencias II, Haya de la Torre y Medina Allende s/n, X5000HUA Córdoba, Argentina
e-mail: nataliap@fcq.unc.edu.ar

C.D. Borsarelli (✉) · V. Rey
Laboratorio de Cinética y Fotoquímica (LACIFO), Centro de Investigaciones y Transferencia de Santiago del Estero (CITSE-CONICET), Universidad Nacional de Santiago del Estero (UNSE), RN9, Km 1125. Villa El Zanjón, CP 4206 Santiago del Estero, Argentina
e-mail: cdborsarelli@gmail.com

E.I. Alarcon et al. (eds.), *Silver Nanoparticle Applications*, Engineering Materials,
DOI 10.1007/978-3-319-11262-6_2

on characteristics such as size, shape and capping-coating. Synthetic approaches for the preparation of AgNP continue to grow as evidenced from the quasi-exponential increase in the number of articles published over the last two decades (Fig. 1).

Generally, the methods used for the preparation of metal nanoparticles can be grouped into two different categories *Top-down* or *Bottom-up*. Breaking a wall down into its components–the bricks, represents the *Top-down* approach, Fig. 2. While building up "the brick" from clay-bearing soil, sand, lime and water would represent *Bottom-up*, Fig. 2. Thus, in nanosciences *Top-down* involves the use of bulk materials and reduce them into nanoparticles by way of physical, chemical or mechanical processes whereas *Bottom-up* requires starting from molecules or atoms to obtain nanoparticles [10].

Top-down fabrication of nanomaterials usually comprise mechanical-energy, high energy lasers, thermal and lithographic methods. Examples of these categories include, but are not limited to, Atomization, Annealing, Arc discharge, Laser ablation, Electron beam evaporation, Radio Frequency (RF) sputtering and Focused ion beam lithography [10].

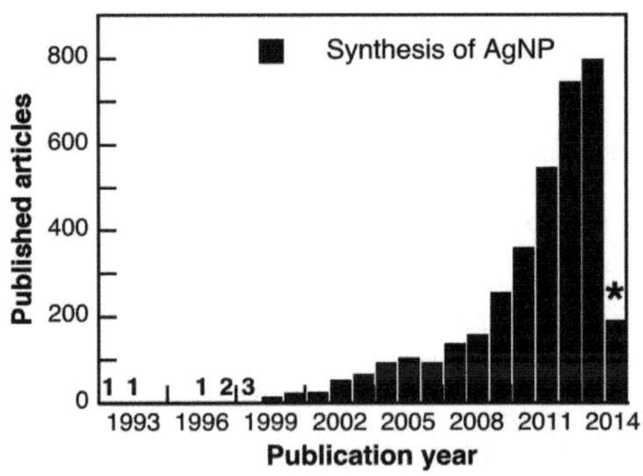

Fig. 1 Representation of the number of research articles published in the period 1992–2014 according to Scopus® containing the term "synthesis of silver nanoparticles" as keyword. Inset numbers indicate (from *left* to *right*) the amount of articles published in 1992, 1993, 1996, 1997 and 1998. The *asterisk* indicates that this result is partial (January–April 2014)

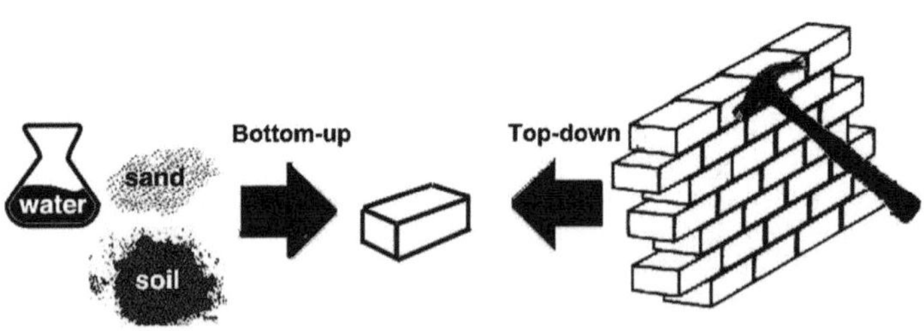

Fig. 2 Illustration of the concepts of *Bottom-up* and *Top-down* methods

Bottom-up production of nanomaterials is divided into the following categories: gaseous phase, liquid phase, solid phase, and biological methods. Among others, chemical vapor deposition and atomic layer deposition belong to the gas-phase methods whereas reduction of metal salts, sol-gel processes, templated synthesis, and electrodeposition correspond to liquid-phase methods [10].

Due to the numerous scientific articles published in the field of synthesis of silver nanoparticles; in this chapter we have focused on providing a rationalized view of some of the available synthetic methods to obtain silver nanoparticles, mostly bottom-up, in liquid phase, excluding biological and microbiological synthesis as reported elsewhere [11, 12].

The main aim of this contribution is to provide guidance when choosing a synthetic method to prepare AgNP for a giving application. Mechanistic insights to understand why some factors would affect the synthetic outcome are also discussed in this chapter.

Although some reviews on synthetic procedures for the preparation of AgNP are reported in literature [13–15], only a few have focused on mechanistic features. In this chapter, we present a systematic review of the mechanism(s) involved in the synthesis of AgNP in the hope "to light up the black box." Further, in the end of this chapter, we have included a summary and a table containing the most commonly used characterization techniques and the information obtained from them.

2 Chemical Reduction

Reduction of the corresponding metal cation represents a straightforward reaction to obtain metal nanoparticles. The key relays on selecting the right parameters that permits control over the synthesis outcome, and so a good understanding on the mechanism is required.

Generally, these reactions are carried out in solution and the product has colloidal characteristics. For this reason, the common term used for the overall phenomenon is co-precipitation, that involves the concurrence of different phenomena; reduction, nucleation, growth, coarsening, and/or agglomeration [16]. The way these processes take place is, in fact, the mechanism of the synthesis.

As for any redox reaction, the values of the standard reduction potentials (E^0) determine the pairs of reactants required for successful chemical conversion. This means that the free energy change in the reaction, ΔG^0, must be negative, or what is equivalent $\Delta E^0 > 0$. Thus, in the case of silver, the relatively large electropositive reduction potential of $Ag^+ \rightarrow Ag^0$ in water ($E^0 = +0.799$ V, [17]) permits the use of several reducing agents e.g., sodium citrate ($E^0 = -0.180$ V, [18]), sodium borohydride ($E^0 = -0.481$ V, [17]), hydrazine ($E^0 = -0.230$ V, [16]) and hydroquinone ($E^0 = -0.699$ V, [19]). Next, we will look at a few examples more in detail.

2.1 Reduction by Citrate Anion

In 1951, Turkevich [20] reported the synthesis of gold nanoparticles in aqueous solution at boiling temperature using sodium citrate to reduce $AuCl_4^-$. Since then, this methodology, known as the Turkevich's method, has been extended to other metals such as the case of silver [3, 21].

Lee and Meisel [3] prepared AgNP in water, for SERS applications, using the described method but in that particular case no insight into the mechanism or full characterization of the AgNP (size and shape) was given. Nevertheless, a few years later, researchers became more interested on elucidating the actual mechanism involved in the whole process in order to gain more knowledge on what parameters really matter and how it will be possible to achieve better reproducibility between batches, and also size and shape control.

From the pioneering studies [3, 20], it is now known that citrate acted both to reduce the metal cation and stabilize the resulting nanoparticles. Also, it was believed that this reactant played a role on determining the growth of the particles. Pillai and Kamat [21] investigated the action of citrate on controlling the size and shape of AgNP. They found that by using the boiling method at different citrate concentrations, AgNP with plasmon maximum absorbance at 420 nm were produced. By increasing the relative concentration of sodium citrate to silver cation i.e. [Citrate]/[Ag^+] from 1 to 5 times, the elapsed time for formation of AgNP was reduced from 40 to 20 min, respectively, indicating that under equimolar conditions a fraction of the Ag^+ was not reduced.

In order to learn more about the function of citrate as stabilizer, SiO_2@AgNP were synthesized using $NaBH_4$ and after the addition of sodium citrate, where the formation of SiO_2-Ag-citrate complex with an association constant of 220 M^{-1} between citrate and the silver colloids was confirmed [21].

In addition, synthesis studies of AgNP by reduction with pulse radiolysis proved that citrate anions act at early stages by complexing Ag_2^+ dimers, and so modulating the particle growth [21]. In fact, the absorbance maximum of the plasmon obtained with this method was found ≈400 nm, a 20 nm blue shifting of the value observed for the AgNP obtained by classical Turkevich's method, indicating that different mechanism of growth particle is operating depending on the reduction method. This interaction had also been observed earlier by Henglein and Giersig [22] in their work on the capping effect of citrate on AgNP prepared by radiolytic reduction.

As a consequence of the slow rate in the citrate reduction method, there is an evident contribution of this reactant to obtain larger AgNP (50–100 nm) [21]. In other words, once the first particle seeds are formed from the Ag^+ reduction by citrate, the remaining anion can complex to the metal surface decreasing the total amount of citrate available in the bulk to further reduce more Ag^+. Thus, fewer new seeds are formed and the initial particles begin to grow via Ostwald ripening, in which the larger particles grow at expense of the smaller ones. Therefore, more time is needed to complete the reduction reaction when using this method. Figure 3 illustrates this process.

Addition of glycerol (40 % V/V), a highly viscous solvent ($\eta \approx$ 1,400 cp [23]), to the aqueous medium reduced polydispersity (≈5 %) and permitted size control

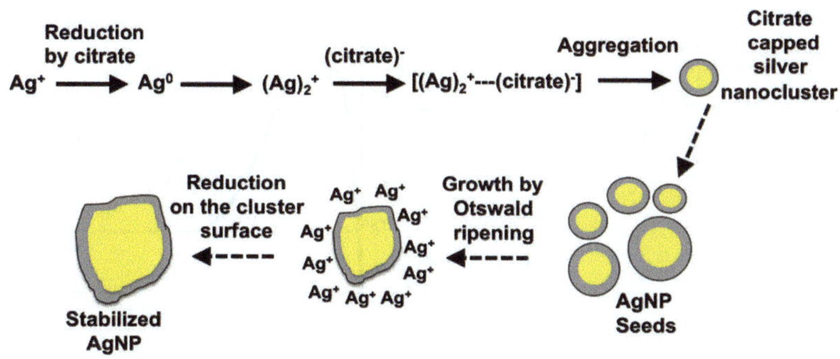

Fig. 3 Representation of the nucleation and growth mechanisms for AgNP obtained by the citrate method according to Ref. [21]

(30 nm) without affecting the spherical shape [24]. Reduction and/or nucleation rates are slower as evidenced by the delayed appearance of the characteristic yellow colour for AgNP, accordingly with expected for diffusion processes in viscous solvents. Although the effect of glycerol on the AgNP synthesis is not fully understood, it is believed that glycerol protects the AgNP against further ripening [24].

The presence of different amounts of NaOH during the synthesis by citrate reduction was found to redirect the reaction to the production of crystalline silver nanowires [25]. These Ag nanostructures were characterized by TEM (observing wires up to 12 μ long), X-ray energy dispersive microanalysis and UV-vis spectroscopy in which a sharp absorption at 370 nm corresponding to the transverse plasmon was observed. The effect of NaOH in the outcome of the synthesis was attributed to interference of hydroxide with the association and capping ability of citrate with silver [25].

2.2 Reduction by $NaBH_4$

First attempts to elucidate the mechanism of AgNP synthesis using sodium borohydride (Eq. 1) as the reducing agent were made by Van Hyning and Zukoski [26]. Following the reaction progress 'in-situ' by UV-vis spectroscopy and 'ex-situ' by Transmission Electron Microscopy (TEM) they were able to infer that the nucleation and growth mechanisms for these nanoparticles do not follow the La Mer's model [27, 28] (Fig. 4), rather it was dependent on colloidal interactions [26]. Recently, Polte et al. [29] proposed a better description of the AgNP formation pathway and the relevant factors to obtain size-controlled AgNP based on a rational design [29].

$$M^{n+} + n\,BH_4^- \rightarrow M^0 + n\,B(OH)_3 + n\,3.5\,H_2 \qquad (1)$$

From time-resolved synchrotron small-angle X-ray scattering (SAXS) and UV-vis spectroscopy measurements combined with TEM characterization, a four-step growth mechanism for the AgNP synthesis by reduction with $NaBH_4$ was proposed (Fig. 5) [29].

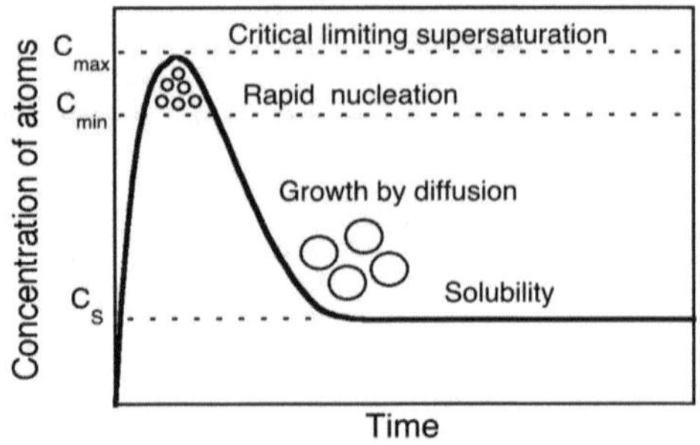

Fig. 4 Representation of La Mer´s model for nucleation and growth [27, 28]. Cs, C_{min} and C_{max} are solubility concentration, minimum and maximum concentration to start nucleation, respectively

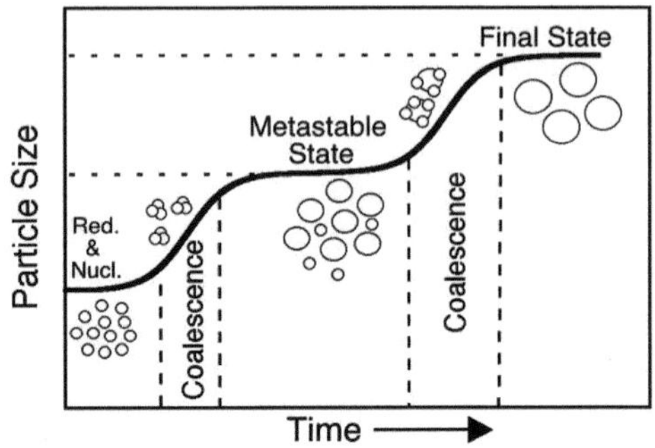

Fig. 5 Illustration of the growth mechanism for AgNP synthesized using $NaBH_4$ as proposed by Polte et al. [29]

The first step involves reduction (<200 ms) of Ag^+ to Ag^0 atoms. These atoms form dimers, trimers, etc. clusters. In a second stage (≈5 s) the clusters coalesce to generate small particles with a radius of 2–3 nm. This step is followed by a metastable state, where the particles maintain a constant size for around 5–10 min. After this period, a second and last coalescence phase takes place (within 30–60 s) to render the final AgNP with an average size of 5–8 nm (Fig. 5) [29]. A detailed study of the metastable state and final coalescence phase by time-resolved SAXS showed that colloidal stability during the intermediate state is enough to prevent further growth due to particle aggregation [30]. Note that each coalescence process implies an aggregation process as a consequence of a decrease in particle stability. Then, it was proposed that the total consumption of BH_4^- is a triggering event for the initiation of the second coalescence phase due to loss of stability plausible by oxidation of the metal surface [30]. Although the reaction stoichiometry with Ag^+ is usually 1:1, BH_4^- is added in excess. Therefore, the extra borohydride anion remains in solution helping to stabilize the formed nanoparticles (electrostatic stabilization), but it will also begin to hydrolyze according to the following simplified Eq. (2), decreasing the available borohydride ion. Hydrolysis is slower compared to the reductive reaction, becoming the dominant process once all the metal cations are reduced.

$$BH_4^- + 4\,H_2O \rightarrow B(OH)_4^- + 4\,H_2 \tag{2}$$

Another phenomenon that takes place is the partial oxidation of the metal nanoparticle surface forming silver oxides. The presence of these oxides decreases the electrostatic stabilization of the AgNP provoking in consequence their aggregation. When there is still BH_4^- left in solution, the oxidation of the particle surface can be reversed, avoiding the aggregation process but once all BH_4^- is consumed, there is green light to start the final coalescence stage until the stable size is obtained [30].

The addition of a bulky stabilizing agent such as polyvinylpyrrolidone (PVP) to the reaction medium does not change the growth mechanism (the final size is indeed the same as without PVP) but affects the duration of each step and decreases the polydispersity (15–20 %) [29].

Based on a better understanding of the growth mechanism, Wuithschick et al. [30] were able to obtain reproducible sizes (between 4 and 8 nm) between different batches by controlling the $BH_4^-/B(OH)_4^-$ ratio, without the addition of any external stabilizing agent. It was found that a proportion of 32.5 % BH_4^- at the beginning is similar to start the synthesis from the metastable state (Fig. 5), and so the four-step mechanism is reduced to a two-step. Consequently, as the growth mechanism is simplified a good control over the final size is gained.

In summary, using a strong reductant, like $NaBH_4$ ($E^0_{red} = -0.481$ V, [17]), results in a faster reduction of Ag^+ than the growth process. The growth mechanism is governed by coalescence closely related to the electrostatic stabilization of the nanopartices. Classical models for nucleation and growth processes (like La Mer´s) do not apply in this case.

2.3 Reduction by Hydroquinone

Oxidation of Hydroquinone (HQ, Scheme 1) involves a two-electron transfer and loss of two protons producing *p*-benzoquinone [19]. According to the above-mentioned redox potential, HQ is able to reduce Ag^+ to obtain AgNP.

Nucleation and growth of AgNP obtained using HQ have been studied by UV-spectroscopy combined with TEM and titration microcalorimetry [31, 32]. Pérez et al. [31] observed a strong dependence on the size and morphology of AgNP with the concentration of HQ at early stages of reaction. A high local concentration of HQ induces more nucleation sites whereas a low local concentration of the reducing agent favors growth over nucleation. Consequently, a protocol involving addition of a concentrated solution of HQ to the reaction medium yields AgNP with smaller sizes (10–30 nm). On the other hand, a protocol in which a diluted solution

Scheme 1 Two-electron oxidation of Hidroquinone to *p*-Benzoquinone

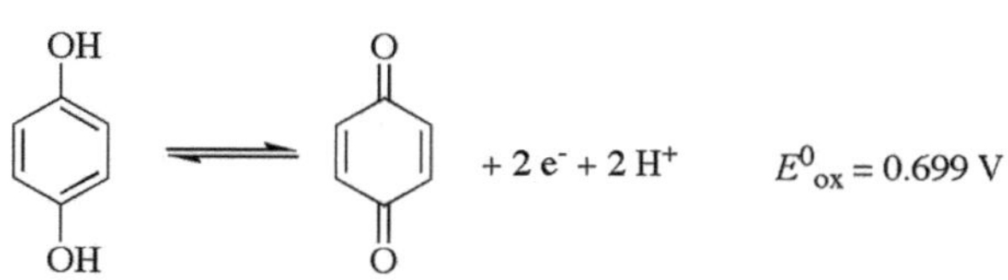

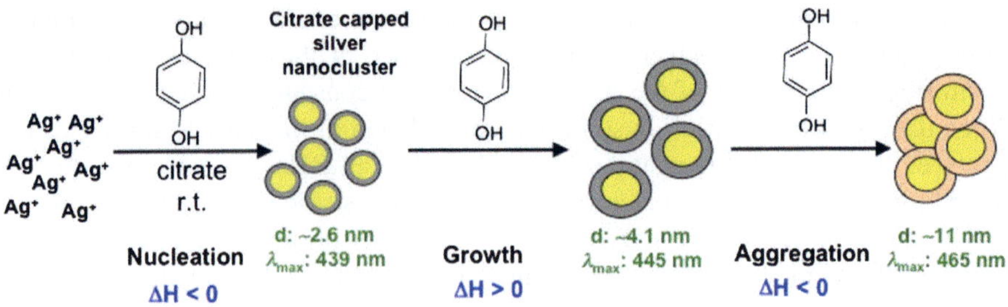

Fig. 6 Illustration of the growth mechanism for AgNP synthesized using HQ at room temperature as proposed in Ref. [32]. Change color in NP shell indicates HQ replaced citrate as capping agent

of HQ is added to the reaction yields larger and more polydisperse AgNP with sizes up to 200 nm. The previous observations are valid for $AgNO_3$ as precursor, where it can be assumed that Ag^+ is completely solvated in solution. Nevertheless, when Ag^+ is complexed as in $[Ag(NH_3)_2]^+$ with the presence of counter ions as Cl^-, the growth is arrested independent on the protocol followed, monodispersity raises and small quasi spherical particles of ~14 nm are predominantly obtained. This effect has been attributed to an unfavorable adsorption of the metallic precursor onto the metal surface, then the growth by further reduction of Ag^+ on the AgNP surface is quenched [31].

Isothermal titration calorimetry (ITC) studies on the formation of AgNP, starting from $AgNO_3$ and hydroquinone in the presence of sodium citrate as stabilizer, reports that the growth mechanism follows a three-step mechanism (Fig. 6). First, reduction of Ag^+ by HQ to get silver clusters (~2.6 nm, λ_{max} = 439 nm) takes place. This nucleation phase is an exothermic (favorable) process that is followed by a growth phase of endothermic nature to reach sizes of ~4.1 nm (λ_{max} = 445 nm) and finally, if aggregation of AgNP occurs the process is again exothermic with final nanostructures around 11 nm in size (λ_{max} = 465 nm) [32]. This latter process occurs with further addition of HQ that is likely to be replacing citrate from the NP surface and thus, decreasing stabilization due to electrostatic repulsion.

2.4 Reduction by Gallic Acid

Reduction of Ag^+ in water at room temperature can also be accomplished using gallic acid (GA) whose oxidation potential is 0.5 V (Scheme 2). Besides the redox potential value, the presence of hydroxyl groups in the benzoic acid structure at determined positions plays an important role in the synthesis of metal nanoparticles. For example, the generation of nanoparticles was not successful when using benzoic acid derivatives substituted by hydroxyl groups at *meta* positions but are obtained when the hydroxyls are located at *orto-* and

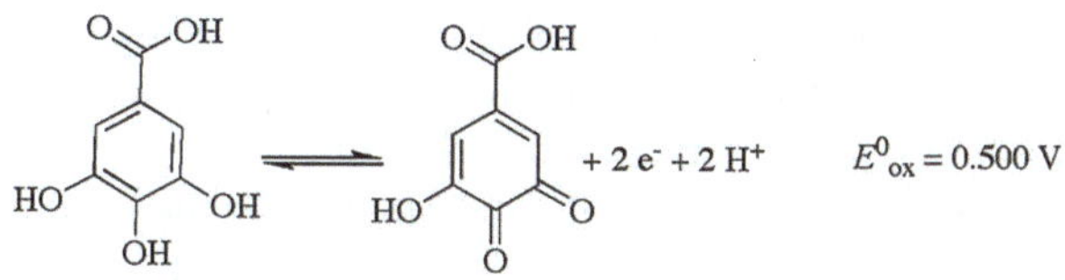

Scheme 2 Two-electron oxidation of gallic acid to the quinone form [33]

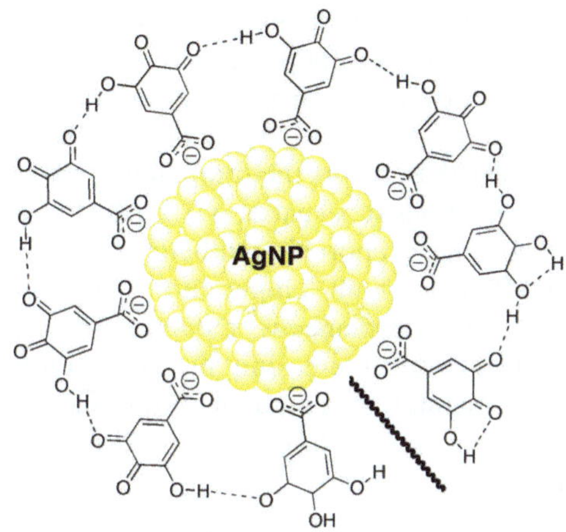

Fig. 7 Illustration of the stabilized capped-AgNP synthesized using GA as proposed in Ref. [33]

para positions [33]. Note that the hydroxyls are the reactive part (Scheme 2) and the carboxylic group acts as the stabilizing part (Fig. 7). The need of having hydroxyls at the mentioned positions can be explained by the formation of a network of the capping molecules around the particle, held by hydrogen bonding interactions between the surface molecules (Fig. 7), that increases the colloidal stability (zeta potential; $\xi = -45$ mV) [33]. Moreover, addition of NaOH is necessary to obtain the silver colloids [33]. Then, the silver species reacting could be Ag_2O that has been recently reported as a good AgNP precursor by thermal decomposition [34].

2.5 Synthesis of AgNP in Organic Solvents

Most reported examples of AgNP obtained by chemical reduction are performed in aqueous media [35–37]. A post-synthesis transfer to an organic solvent is usually difficult due to aggregation processes. Nevertheless, the synthesis of metal nanoparticles in organic solvents has some advantages such as high yield and narrower size distribution, with the additional advantage that in some cases the solvent itself can act as reducing agent to obtain AgNP [14].

For example, Pastoriza-Santos and Liz-Marzan [38] proposed a synthetic route to obtain AgNP spheres (≈6–20 nm) from the spontaneous reduction of silver nitrate by dimethylformamide (DMF) and using 3-(aminopropyl)trimethoxysilane (APS) to stabilize the particles. The overall reaction displayed in Eq. 3 shows the formation of a carbamic acid that together with the resulting acidic environment promote the deposition of silica shells onto the formed particles [38]:

$$HCON(CH_3)_2 + 2\,Ag^+ + H_2O \rightarrow 2\,Ag^0 + (CH_3)_2NCOOH + 2\,H^+ \qquad (3)$$

Reaction rate, size and monodispersity of the nanostructures can be controlled by variation of temperature from room to boiling temperature. Increasing the temperature accelerates the silver reduction favoring silica polymerization directly onto the APS-coated silver particles surface. Thus, at 156 °C APS-capped AgNP around 19 nm are obtained [38]. Replacing the stabilizing agent APS by polyvinylpyrrolidone (PVP) favors the formation of nanoprisms by choosing the right experimental conditions (concentration of reactants and reflux time) [39].

Another reported synthetic route was performed in dimethyl sulfoxide (DMSO) at room temperature. Quasi-spherical AgNP (4.4 nm) were obtained from the spontaneous reduction of silver 2-ethylhexanoate [Ag(ethex)] by DMSO [40]. Increasing the temperature accelerates the reaction. Based on experimental data and ab initio calculations a mechanism for this reaction has been proposed and it is shown as follows (R is—$(C_2H_5)CH(CH_2)_3CH_3$ in Eqs. 4–7):

$$[AgOOCR] + DMSO \rightleftharpoons [AgOOCR \bullet DMSO] \qquad (4)$$

$$[AgOOCR \bullet DMSO] \rightarrow Ag^0 + [DMSO^\bullet]^+ + [:OOCR]^- \qquad (5)$$

$$[AgOOCR] + [DMSO^\bullet]^+ \rightarrow Ag^0 + [:OOCR]^- + [DMSO]^{2+} \qquad (6)$$

$$[DMSO]^{2+} + H_2O \rightarrow DMSO_2 + 2H^+ \qquad (7)$$

The overall reaction can be written as follows (Eq. 8).

$$2[AgOOCR] + DMSO + H_2O \rightarrow 2Ag^0 + DMSO + [:OOCR]^- \qquad (8)$$

Note that in the first step (Eq. 4) the formation of a complex between Ag(ethex) and DMSO is proposed which was supported by testing that the reaction does not occur with other silver salts such as $AgNO_3$ and $AgClO_4$ otherwise 2-ethylhexanoate is present in the reaction medium. In addition to that, ab initio calculations showed that the complex [Ag(DMSO)ethex] has an adequate LUMO energy and low energy difference between HOMO-LUMO gap what is consistent with a good precursor for reduction [40].

In order to overcome the instability of the synthesized AgNP dispersions by this way, it is necessary to use capping molecules to gain stability. The best capping agent used was sodium citrate (see Sect. 1).

A deeper analysis of the synthesis of AgNP in organic solvents with focus on rationalizing the nucleation process to engineer the final outcome can be found in a tutorial review by Sun [14].

2.6 Polyol Method and Shape-Controlled Synthesis

Although the polyol method can also be considered as a case of synthesis in organic solvents [14, 41, 42], the versatility that provides to obtain AgNP with different morphologies makes this method relevant to be analyzed separately.

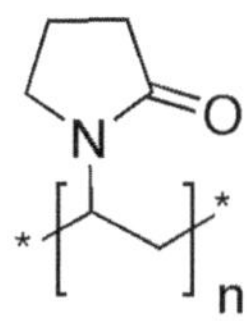

PVP

Briefly, polyol synthesis involves the reduction of a metal salt used as precursor by a polyol, usually ethylene glycol (EG) at an elevated temperature (≈160 °C for EG), and in order to prevent agglomeration of the particles a capping agent commonly used is PVP [4, 43–45]. EG (as other polyols) probably oxidizes to aldehyde species with an oxidation potential of 1.65 eV [46]. Reduction power of EG is markedly dependent on the temperature, giving the ability to control the nucleation and growth processes by choosing the reaction temperature [47]. Another advantage is that the polyol can serve as both solvent and reducing agent.

Before analyzing the reasons for shape control during the synthesis by the polyol method, it is important to note the critical role of PVP in the stability, size and shape uniformity of AgNP. From infrared (IR) and X-ray photolectron spectroscopy (XPS) it was inferred that the oxygen and nitrogen atoms of PVP can promote the adsorption of this polymer chains onto the metal surface [36]. In addition, PVP interacts more strongly with silver atoms located in the {100} facets than those on the {111} (lower energy facets) [42].

La Mer's model (Fig. 4) seems to explain at some extent the nucleation and growth mechanisms [14, 36], taking place in the polyol method though it still represents an oversimplified view of the nucleation process [14]. That said, we now summarize different examples of shape control: nanocubes, nanowires and nanospheres.

TEM studies performed on small silver nanoparticles demonstrated that there is a good probability that morphology fluctuates between the kinetically stable single crystal (sc) and the thermodynamically stable twinned (tw) particles during earlier stages of growth [44]. This fact supports a model where morphology in small particles (<5 nm) is exchanged between sc and tw induced by thermal fluctuations, with multiply twinned decahedra (MTP) being the most abundant shape. Thus, it is understood that a common NP synthesis outcome results to be a mixture of single crystals and twinned particles as shown in Fig. 8 [44]. It also has been proposed that sc seeds lead to obtaining truncated cubes and tetrahedrons as final shapes while tw seeds conducts to rods, wires and spheres depending on the reaction conditions [47].

Experimental conditions that favor fast nucleation and fast growth like a high $AgNO_3$ concentration (0.125–0.25 M) and low PVP to Ag^+ ratio (≈1.5), promote the formation of nanocubes as the possibilities to get twinned particles diminished. Then, the most abundant structures between the pristine particles are single crystals that follow their growth assisted by PVP to reach the final shape of truncated cube. Figure 9

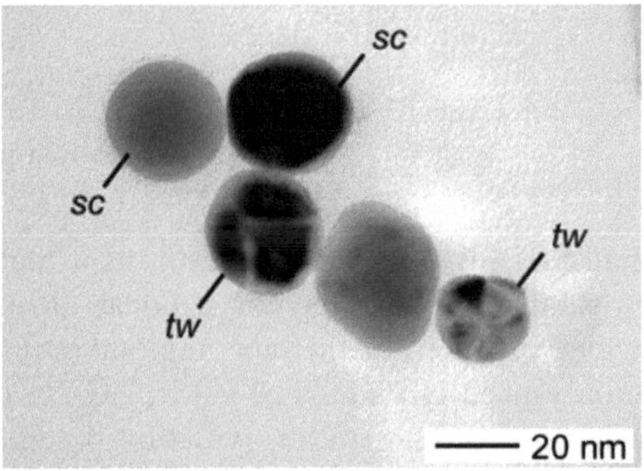

Fig. 8 TEM images showing a mixture of single crystal and twinned morphologies after 2 h reaction for AgNP synthesized by the polyol method. Reprinted with permission from Ref. [44]. Copyright © 2004 American Chemical Society

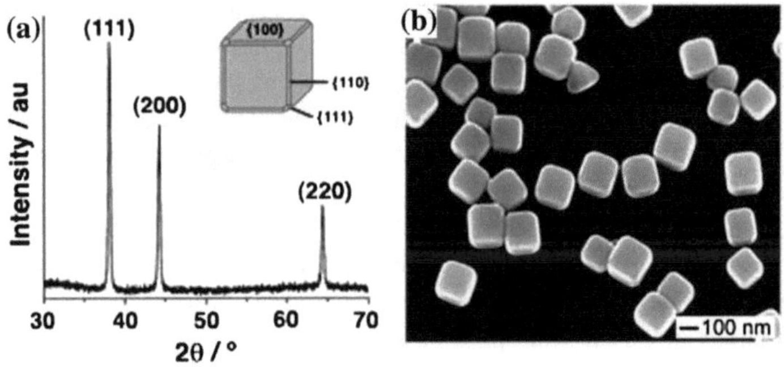

Fig. 9 **a** X-ray diffraction (XRD) pattern obtained from silver nanocubes, synthesized by the polyol method from $AgNO_3$ (0.25 M) with a PVP to Ag^+ ratio of 1.5, deposited on a glass substrate. Peaks are assigned to diffraction from the {111}, {200} and {220} planes of silver. Inset shows the drawing of one cube indicating the corresponding crystallographic planes. **b** SEM image of the same silver nanocubes showing the slightly truncated corners and edges of each cube. Reprinted with permission from Ref. [47]. Copyright © 2005 WILEY-VCH Verlag GmbH&Co. KGaA, Weinheim

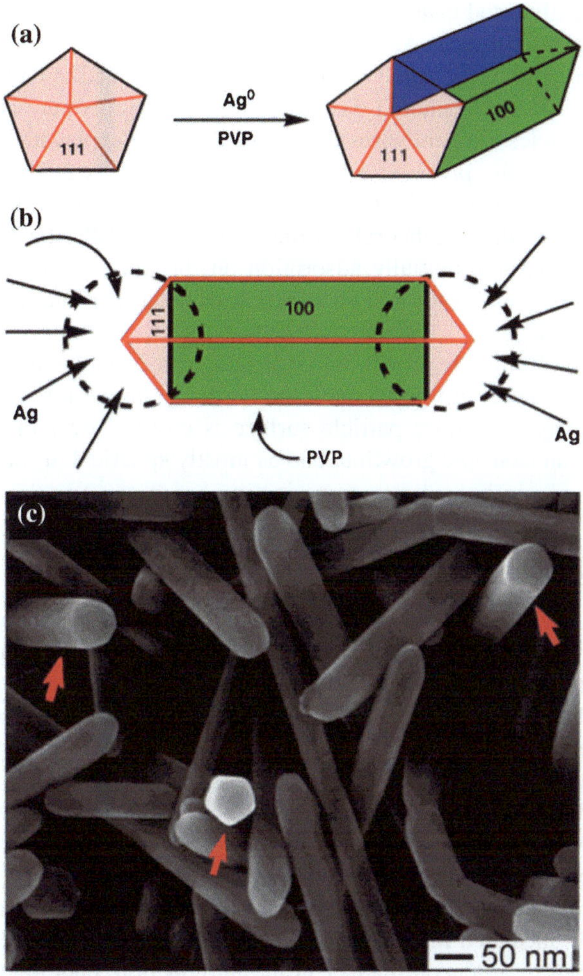

Fig. 10 **a** Representation of the evolution of a nanorod from a decahedral silver particle with the assistance of PVP by the polyol method. The ends of this nanorod are terminated by {111} facets, and the side surfaces are bounded by {100} facets. The green color indicates the preferential adsorption of PVP on the {100} facets, and the pinkish color stands for the weaker interaction with the {111} facets. The red lines on the end surfaces represent the twin boundaries that can serve as active sites for the addition of silver atoms. The plane marked in blue shows one of the five twin planes that can serve as the internal confinement for the evolution of nanorods from MTP. **b** Representation of the diffusion of silver atoms toward the two ends of a nanorod, with the side surfaces completely passivated by PVP. This drawing shows a projection perpendicular to one of the five side facets of a nanorod, and the arrows represent the diffusion fluxes of silver atoms. **c** SEM image of silver nanowires obtained by the polyol method. The red arrows indicate cross-sections of the nanowires showing pentagonal cross-sections. **a**–**c** were adapted with permission from Ref. [42]. Copyright © 2003 American Chemical Society

shows the silver truncated nanocubes (average edge length: 175 nm) obtained by the described polyol method from $AgNO_3$ (0.25 M) [47].

On the other hand, if the $AgNO_3$ concentration is reduced to 0.085 M while the PVP:Ag^+ ratio is kept constant (1.5) and the reaction temperature is above 110 °C the resulting particles are nanowires. The most accepted explanation for this result is that these conditions promote a high yield in MTP (decahedral) during early growth stages. Then, new Ag^0 are more likely to crystallize on the highly reactive twin defects of the decahedral. In the presence of PVP, the {100} sides are passivate due to its preferentially adsorption on this facets. For that reason, further growth takes place in the {111} direction to reach the final nanowires shape (Fig. 10) [42, 46, 47].

In order to overcome the growth in a particular direction due to passivation of a specific facet, it is necessary to increase the PVP:Ag^+ molar ratio more than three times. In that way, the entire particle surface is covered by PVP and seeds are prone to follow an isotropic growth to end as mostly spherical in shape [47].

3 Light-Assisted Methods

The first reports on the effect of light on the transformation of some metal salts were written in the 18th century, when the German scientist Johann Schulze (1687–1744) discovered that silver salts darken by irradiation of light, initiating the development of photography, although the concept of nanomaterials had to wait almost 200 years.

In the last decades, light-assisted methods for NP preparation have been largely developed as well, basically due to the versatility and selectivity of photochemical reactions to generate in situ the reducing species with high spatial resolution and almost no modification of the surrounding media. The fundamentals and applications of different light mediated methodologies have been recently reviewed [13, 15, 48]. As we mentioned in a general manner in the introduction, photo-induced synthetic strategies can be also categorized into top-down (photophysical methods) and bottom-up (photochemical methods) approaches.

3.1 Photophysical Methods

AgNP are prepared via subdivision of bulk silver metal usually by laser ablation methods [49–51]. As example, the influence of anionic surfactant molecules $C_nH_{2n+1}SO_4Na$ ($n = 8, 10, 12, 16$) on the stability of AgNP produced by laser ablation of a metal silver plate with a pulsed Nd-YAG laser operating at 532 nm (10 ns full width at half maximum (fwhm), <90 mJ/pulse, 10 Hz) [49]. The stability of the AgNP capped with the anionic surfactant was modulated by the surfactant coverage and the charge state on the nanoparticle surface. The AgNP tend to aggregate when the coverage degree is less than unity, while they are very stable when the particle surface is covered with a double layer of the surfactant due to the electrostatic repulsion forces. Moreover, more efficient stabilization was obtained by surfactant with a longer hydrocarbon chain due to larger hydrophobic interactions between them [49].

Additionally, the effect of sodium chloride (NaCl) on the morphology and stability of AgNP generated by laser ablation (Nd:YAG 1,064 nm) was analyzed [50, 51]. The increment in NaCl concentration up to 5 mM in aqueous solutions produced AgNP with average size of ~26 nm, but with a continuous decrease of the full with at the half maximum (fwhm) of the plasmon absorbance band centered at 400 nm, together with the simultaneous increase of the absorbance signal. However, above that NaCl concentration the absorbance of the plasmon was smaller and the fwhm value slightly increased. The salt effect was inherent with the growth phase of the AgNP during the ablation of the bulk silver surface, since no effect on the plasmon band was observed when NaCl was added to AgNP in water previously prepared by laser-ablation. It was concluded that the presence of Cl^- ions provides reduction of nanoparticles size and increases the efficiency of formation of AgNP by laser irradiation of the silver target immersed in water. However, the long-term stability (i.e. >20 days) of the AgNP was reduced by the Cl^- ions [50].

In another case, stable suspensions of AgNP were prepared by laser-ablation of a highly pure (99.99 %) silicon (Si) target immersed in aqueous solution of $AgNO_3$ as silver source [52]. The third-harmonic ($\lambda = 355$ nm) beam of a Q-switched Nd:YAG laser was focused on the Si target and the effects of the salt concentration and irradiation time were evaluated. The maximum plasmon absorbance at 400 nm was obtained using 0.125 mM $AgNO_3$ solutions, with the formation of spherical AgNP with average diameter of ~11 nm. Above this concentration, the plasmon absorbance was less intense, red-shifted and broader indicating that under these salt concentrations AgNP with larger diameter and a broad size distribution were formed [52].

3.2 Photochemical Methods

Regarding to the photochemical methods for metal NP preparation (bottom-up approach), they are based on the reduction of the metal cation M^{n+} to M^0 either by direct or indirect (photosensitized) photolysis.

Direct photoreduction has been established as an important technique for metal NP synthesis, where M^0 is formed through the direct excitation of a metal source, normally a salt. Due to the advantage of being free from reducing agents, it has been widely employed in the various mediums including polymer films, glasses, cells, etc. [15].

Direct photoreduction of silver perchlorate ($AgClO_4$) in aqueous and alcoholic solutions by irradiation with UV-light at 254 nm was investigated in the 70s by Hada et al. [53]. The photoreduction mechanism is based on the electron-transfer from a solvent molecule to the electronically excited state of Ag^+ to form Ag^0 (Eqs. 9–12):

$$Ag^+ + H_2O \xrightarrow{h\nu} Ag^0 + H^+ + HO^\bullet \tag{9}$$

$$Ag^+ + RCH_2OH \xrightarrow{h\nu} Ag^0 + H^+ + R\dot{C}HOH \tag{10}$$

$$Ag^+ + R\dot{C}HOH \xrightarrow{h\nu} Ag^0 + H^+ + RCHO \tag{11}$$

$$nAg^0 \rightarrow AgNP \tag{12}$$

The reactivity, and therefore the quantum yield of formation of AgNP, was strongly increased in the presence of α-alcohols, due to the easier abstraction of the labile α-hydrogen. The formed radical also reduced Ag^+ (Eq. 11) accelerating the production of AgNP [53].

Long term stable (~6 months) AgNP were obtained by photoreduction of $[Ag(NH_3)_2]^+$ aqueous solution with PVP as both reducing and stabilizing agent [54]. Following the increment in the surface plasmon absorption maxima at 420 nm the formation of AgNP was monitored. It was found that the plasmon absorption intensity increased with PVP concentration as well as the maximum absorption wavelength (λ_{max}) was blue shifted. Transmission electron microscopy (TEM) showed the resultant particles were 4–6 nm in size with uniform particle size distribution. Huang et al. [55] also reported the synthesis of AgNP employing the photoreduction of $AgNO_3$ solutions with 254 nm UV irradiation in the presence of PVP as protecting agent.

Finally, a marked interest in UV photoreduction of silver salts embedded in polymeric films and/or inorganic hydrogels for in situ generation of AgNP exists [56–58]. Fast generation of AgNP was observed when fibers of polymer blends of poly(vinyl alcohol) (PVA) and poly(acrylic acid) (PAA) crosslinked with DMSO were irradiated with 350 nm light. Initially, small AgNP clusters were formed, which were stable in the dark but transformed into larger metal particles upon further illumination [56]. The formation of AgNP by UV photoreduction of $AgNO_3$ in layered laponite suspensions without any addition of reduction agent or heat treatment. Relatively larger sized (40–110 nm) AgNP were obtained, but with extremely long-term stability, opening the application of the hydrogel as antibacterial agent [57].

Thin films of PAA embedded with AgNP prepared by UV photoreduction exhibited cyclically changeable optical absorbance properties (spectral shift) during variation of pH of aqueous medium [58]. The phenomenon was attributed to conformational changes in the polymer matrix induced by pH changes, which leads to variation in the 3D configuration of the AgNP ensemble by reversible modification of the inter-particle distance during swelling and shrinking processes within the PAA matrix (Fig. 11). These changes result in the reversible blue-red shifting of the

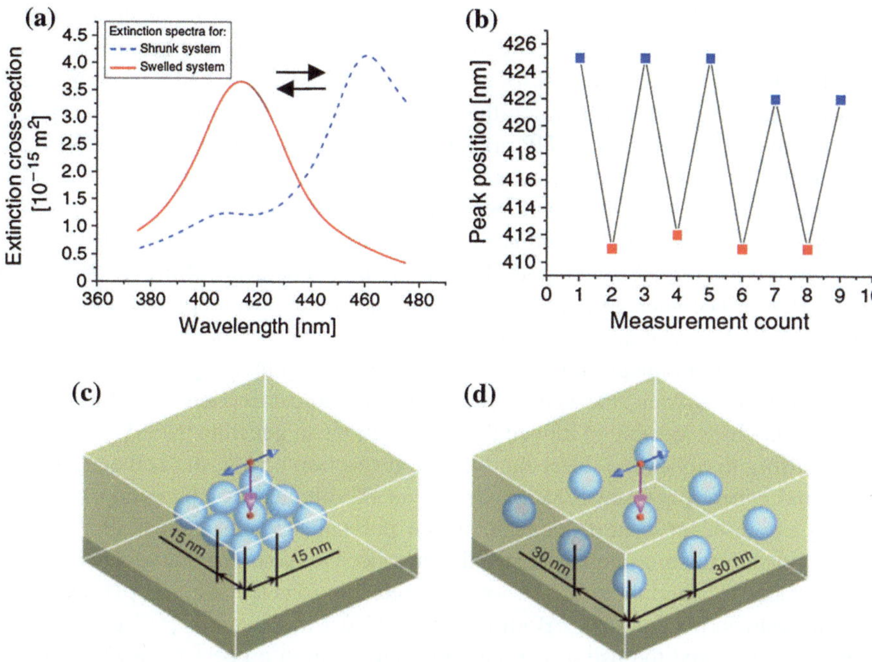

Fig. 11 **a** Electronic absorption spectra of AgNP obtained by direct photoreduction in PAA hydrogel film wet with 0.1 M H_2SO_4 (swelled system) and with deionized water (shrunk system). **b** Absorbance spectrum peak positions after 10 cycles of alternating immersion where red and blue squares indicate the swelled and shrunk systems, respectively. (**c**–**d**) Models of the shrunken (**c**) and swollen (**d**) states of the PAA hydrogel containing AgNP. Adapted with permission from Ref. [58]. Copyright © 2012 CSIRO Publishing dx

plasmon band absorption maximum, with the possibility of using this "molecular memory" in the development of optoelectronic devices and sensors [58].

As despited above, the general mechanism of NP formation by direct photoreduction of the metal cation is usually by a solvent-assisted disproportionation process that depends on the nature of the precursor metal. However, in most cases, the use of UV excitation source is necessary in direct photoreduction since most of metal cations and/or metal salts only absorbs in this spectral region. This issue can represent a very important experimental limitation, since UV-light sources are more expensive and not always easily available. Moreover, many molecules used as stabilizing agents can also absorb in the UV, acting as inner filter of the excitation beam.

Nevertheless, this limitation can be overcome by photosensitized synthesis of NP. In this case a sensitizer molecule (organic dye, aromatic ketones, polyatomic anions) that can absorb UVA-visible light (320–700 nm) generates reducing intermediates (free radicals, anions, solvated electrons, etc.) that react with the M^{n+} coexisting in the solution to form the M^0 [15].

Photosensitized methods are based either on the formation of reactive excited states or the photogeneration of organic radical species from the sensitizer molecule [15]. In this chapter we will focus on the photosensitized generation of AgNP mainly by ketyl radical formation either by hydrogen abstraction or bond cleavage as a very useful strategy for AgNP preparation by indirect photolysis [15, 59].

Typically, in the former case, an aliphatic or aromatic ketone, such as acetone, acetophenone or benzophenone, is excited with UVA/UVB light ($300 < \lambda$ (nm) < 360) to the singlet excited state, in which by efficient intersystems crossing decays to the long-lived triplet state. Due to the n,π* character of the ketone triplet excited state it becomes an excellent hydrogen abstractor from appropriated donors (e.g. alcohols, benzylic and allylic hydrogens, some hydrocarbons, etc.) to produce the respective ketyl radical. In the case of alcohols, which in many cases can be used as solvent or co-solvent, the ketyl radical and the radical derived from the alcohol are formed, which are able to reduce the metal cations to generate the NP.

According to Scheme 3, after the formation reaction of M^0, the ketone is regenerated and a new photosensitization cycle can be initiated by absorption of a UV photon, a significant issue to consider when polyvalent metal cations are reduced since ketyl radicals are one-electron reducing species. Note that the global reaction also produces H^+, and therefore in non-buffered solutions the solution pH decreases to acidic values (pH ~ 3) [59]. Both Majima´s [15] and Scaiano's [59] groups have extensively explored the use of ketone photosensitization process for AgNP synthesis under different conditions and applications. As example, since the primary precursor is the excited triplet state of the ketone, the effect of magnetic field on the formation of AgNP was evaluated [60]. In this case, photoreduction of benzophenone in sodium dodecyl sulphate (SDS) micellar solutions in the presence of Ag^+ produced a very rapid and efficient formation of AgNP, and the process was modulated by external magnetic field by splitting of triplet sublevels as a result of Zeeman effect, even under weak and mild field intensities (up to 120 mT). The amount of AgNP produced was increased with the intensity of the magnetic field with slight modifications in the plasmon absorbance band position and increment of the average particle size from ~4 to 8 nm. Therefore, it was concluded that weak magnetic fields, such as those produced by magnetic stirring bars (<10 mT), can modulate the formation of AgNP by benzophenone photoreduction [60].

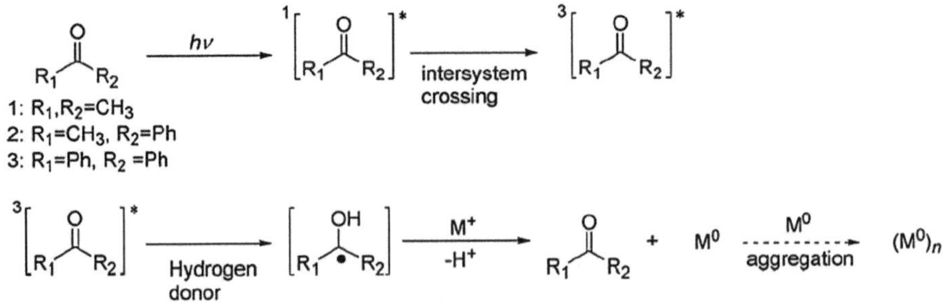

Scheme 3 Sensitization mechanism of ketone/hydrogen donor system used for nanoparticle formation. Reproduced with permission from Ref. [15]. Copyright © 2009 Elsevier

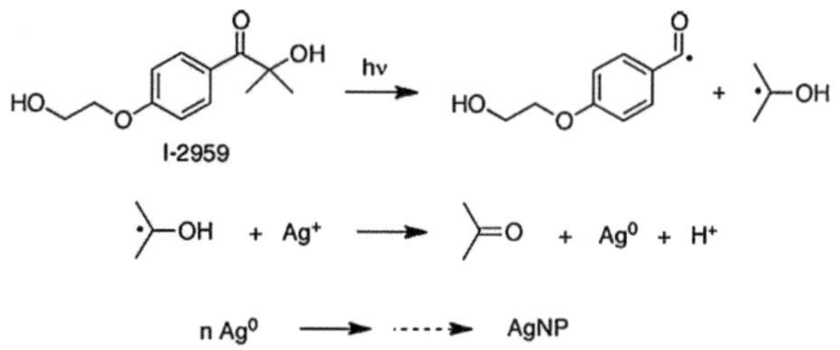

Scheme 4 Mechanism for AgNP formation starting from the photoinduced bond cleavage of Irgacure-2959. Reproduced with permission from Ref. [61]. Copyright © 2013 Springer

Other source of ketyl radical for metal reduction is given by photoinduced bond cleavage reactions of α-hydroxy ketones (benzoins), such as the water-soluble 2-hydroxy-4′-(2-hydroxy-ethoxy)-2-methyl-propiophenone (Irgacure-2959®, I-2959, Scheme 4).

The photochemistry of α-hydroxy ketones has been well characterized and the homolytic rupture of the α-cleavage bond favored by alkylation of the 2-hydroxy group occurs in the picosecond time scale after excitation [62]. The case of I-2959 is remarkable since is commercially available, water-soluble, and produces ketyl radicals with a quantum yield of 0.29 in fast photocleavage from a triplet state with a lifetime of just 11 ns [62]. Scaiano's group, principally, has extensively established the use of this compound as photosensitizer for AgNP preparation in water and organic solvents [48, 61, 63–66]. The methodology is clean, fast, and also allows real-time kinetics monitoring by UV-vis spectroscopy using modular CCD spectrometer devices, as shown in Fig. 12 for the formation of AgNP upon steady-state irradiation at 320 (±10 nm) of 0.2 mM $AgNO_3$ and 0.2 mM I-2959 in the presence of 1 mM sodium citrate aqueous solutions, together with the TEM image of the final AgNP (average diameter ~10 nm) and the yellowish color of the solution due to the strong plasmon absorbance at 404 nm (courtesy of V. Rey, Universidad Nacional de Santiago del Estero).

Finally, it should be noticed that the final fate of the substituted benzoyl radical formed after homolytic bond scission in the first reaction of Scheme 4, is the formation of the corresponding carboxylic acid 4-(2-hydroxyethoxy) benzoic acid (HEBA) that, through mild binding, contributes to the stability of "naked" (without extra stabilizers) metal nanoparticles [48].

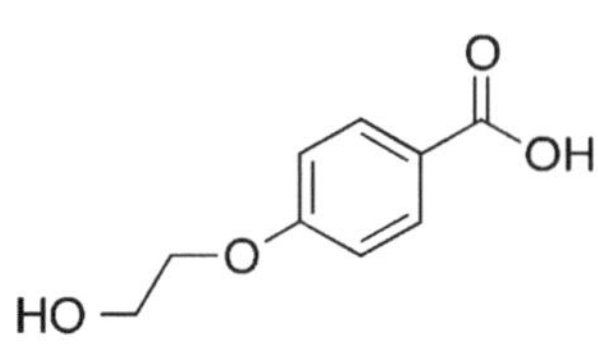

HEBA

At this point, it is worth commenting on the noticeable effect of light on the morphology of the AgNP during the growth stage of Ag^0 seeds [13, 36, 51, 64, 67, 68]. In general terms, the observed effect is that depending on the excitation wavelength (~400–700 nm) from lasers, lamps or light emitting diodes (LEDs) illumination sources during the ripening of small spherical AgNP (<5 nm) seeds,

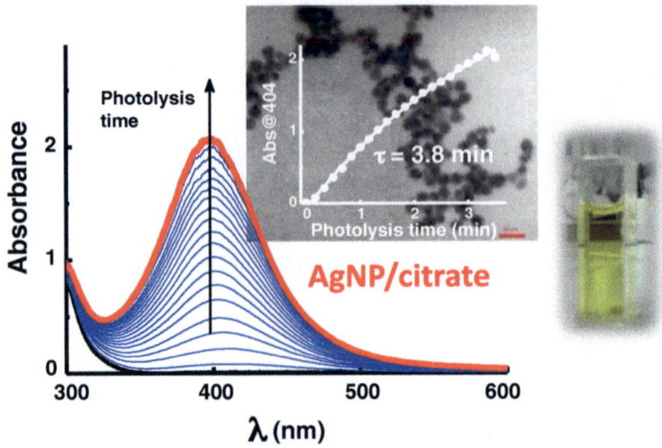

Fig. 12 Absorption spectra of AgNP produced by using I-2959 as photosensitizer showing the formation kinetics. Inset shows the absorbance dependence over photolysis time monitored at 404 nm together with a TEM image and regular photograph of the final AgNP solution. Reaction performed with continuous UV light irradiation in the presence of citrate anions as stabilizer. Courtesy of Dr. Valentina Rey, LACIFO-UNSE, Argentina

different nanostructures with shape and size control, including larger spheres, dodecahedra, nanoplates, and nanorods are photogenerated (Fig. 13).

This striking size and shape control on AgNP formation is associated with the fact that the electromagnetic (EM) field generated in the vicinity of AgNP is extremely high in comparison with other metals.

In summary, light-assisted preparation of AgNP is a very promising field because of the large versatility, selectivity and relatively easier control on shape and size of matured AgNP without need of harsh reaction conditions, such as boiling temperature, etc. These mild reaction conditions also allow NP preparation

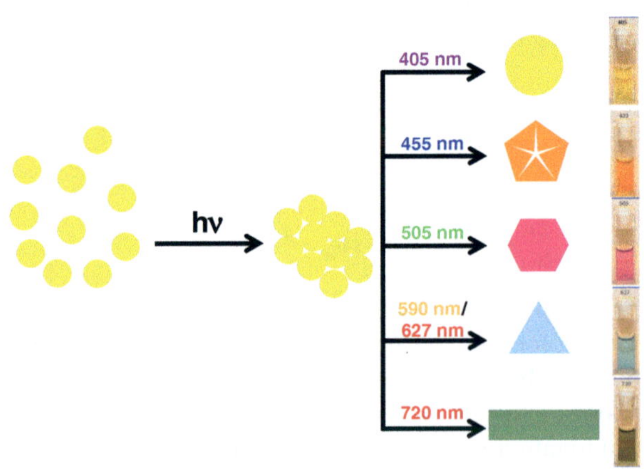

Fig. 13 Representation of the photoinduced aggregation/coalescence mechanism that leads to morphology transformation of AgNP seeds according to Ref. [68]

with different labile or biological protecting agents, such as proteins. The current commercial availability of UVA-vis high power LEDs can allow inexpensive preparation of matured AgNP with modulated shape and size for different desirable applications.

4 Electrochemical Methods

Considering that the primary step for AgNP preparation is the one-electron reduction of Ag^+ cation, electrochemical reduction to generate metallic particles seems to be a suitable choice method [69–79].

Synthesis of AgNP by electrochemical reduction was presented by Rodríguez-Sánchez et al. [69], based on the dissolution of a sacrificial silver sheet as the anode (counter electrode), and a sized platinum (Pt) or aluminum sheet was used as cathode (working electrode) and containing tetrabutylammonium salts (TBA bromide or TBA acetate) in acetonitrile as inert solvent. However, AgNP were only obtained using Pt as working electrode. The spherical particles were further characterized by TEM and UV-vis spectroscopy. The increment of the current density was analyzed, observing a size reduction of AgNP from ~7 to 2 nm accompanied by a red shift of the broad plasmon absorption band from 422 to 445 nm. It was observed an oscillatory behaviour by studying various electrochemical parameters and reaction conditions. Characterization by UV-vis spectroscopy showed that the absorption bandwidth of the plasmon band depends heavily on the size and interaction with the surrounding medium, and an autocatalytic stage in the synthesis mechanism was proposed.

Electrochemical synthesis of spherical AgNP (<7 nm) in aqueous media were obtained by using a highly oriented pyrolytic graphite (HOPG) surface modified with a 4-aminophenyl monolayer as counter electrode (Fig. 14) [70]. Changing experimental conditions, such as immersion times, immersion period, and pulse potential time, controlled both the size and number density of AgNP.

Electroreduction of 5 mM $AgNO_3$ aqueous solutions using a rotating platinum cathode and PVP as stabilizer produces nanospheres with narrow size distribution (~20 nm, SD: 14 nm) which is improved by addition of a surfactant like sodium dodecyl benzene sulfonate (SDBS) (~10 nm, SD: 2 nm) [73]. The dual role of PVP in promoting metal nucleation and circumventing agglomeration of metallic particles allows the formation of small sized metal nanoparticles. Furthermore, by using a rotating platinum cathode, the effective transferring of the electrochemically synthesized NP from the cathode vicinity to the bulk solution is favored, avoiding the occurrence of flocculates in the vicinity of the cathode, and ensuring the monodispersity of the particles [73].

Silver colloidal nanospheres have also been synthesized electrochemically assisted with ultrasonication at 45 °C in a complex media consisting of cyclohexane (100 μL), acetone (100 μL), tetraoctylammonium bromide (0.02 g), hexadecyltrimethylammonium bromide (0.1 g), and ultra-pure water (10 ml) [75]. The average

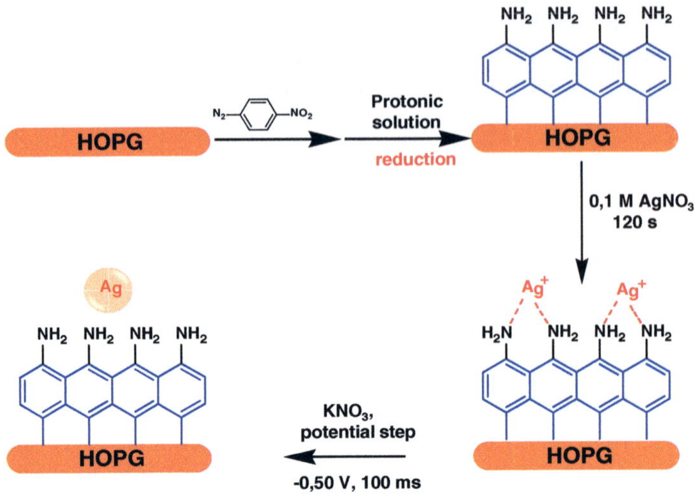

Fig. 14 Illustration of the electrochemical preparation of AgNP on 4-aminophenyl modified HOPG surface as described in Ref. [70]

diameter of the AgNP synthesized in this way is about 12 nm, as determined by TEM. The UV-vis absorption spectrum of fresh prepared AgNP showed a sharp absorption band at 410 nm. However, after some hours of aging of the solution, a second absorption maximum at 470 nm was noticed indicating particle aggregation and clustering. These two absorption bands were assigned to the transverse and longitudinal surface plasmon resonance, respectively. These authors also analyzed the fluorescence emission of the sample as a function of the excitation wavelength, and a continuous red-shift of the emission maximum from 540 to 565 m was observed when the excitation wavelength changes from 380 to 440 nm. Conversely, UV excitation between 230 and 280 nm, yielded emission spectra with maxima fixed about 315 nm. These results suggest that these electrochemically generated AgNP show dual fluorescence emission, where the shorter emission at 315 nm is due to local field enhancement by relaxation of the excited electron and recombination of electrons and holes, while the longer wavelength red shifting bands with the increasing excitation wavelength results from electron interband transitions, see Chap. "Silver Nanoparticles: From Bulk Material to Colloidal Nanoparticles".

Size and shape control of AgNP has been shown by Ueda et al. and Plieth et al. [72, 80] using a double pulse electroreduction method, which is based on an extremely short nucleation pulse of high cathodic polarization followed by a much longer growth pulse at low cathodic over-voltage (Fig. 15). The ideal model situation is where nucleation only occurs within the first pulse and exclusive particle growth in the second pulse. The high cathodic amplitude of the first pulse is necessary in order to initiate nucleation. Using this method, the conflict between both optimal conditions for nucleation and growth is partially defused. This is due to the amount of small seeds additionally nucleated at the higher polarization and

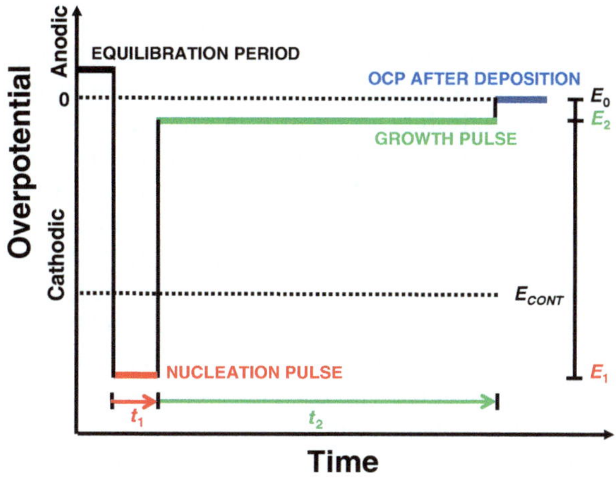

Fig. 15 Representation of the potentiostatic double-pulse method as described in Ref. [72]

resolved as soon as the potential is switched over to the lower polarization of the growth pulse [72].

A deposit consisting of AgNP has been obtained by both potentiostatic and galvanostatic methods in non-aqueous solution of $NaNO_3$ in ethanol (C_2H_5OH) as investigated by cyclic voltammetry and chronoamperometry [76]. The proposed mechanism assumes that both anodic dissolution of Ag^0 and its reduction to metallic state proceed during polarization in ethanol according to the following mechanism (Eqs. 13–17) initiated by the Ag^0 one-electron reaction at the surface of the working electrode and solubilization of Ag^+ by C_2H_5OH,

$$Ag^0 + C_2H_5OH \rightarrow \left[Ag(C_2H_5OH)\right]^+ + e^- \quad E > 0.4\ V \tag{13}$$

which rapidly decomposes to Ag_2O and ethyl ether (Eq. 14).

$$2\left[Ag(C_2H_5OH)\right]^+ \rightarrow Ag_2O + C_2H_5OC_2H_5 \tag{14}$$

The formation and deposition of Ag_2O can partially block the anode. However, at $E > 0.6$ V, the oxidation of ethanol occurs forming ethanal (CH_3CHO)

$$C_2H_5OH \rightarrow CH_3CHO + 2H^+ + 2e^- \tag{15}$$

In turn, the aldehyde reduces Ag_2O easily forming Ag atoms (Eq. 16) and the $Ag(C_2H_5OH)^+$ is reduced in parallel on the cathode (Eq. 17).

$$CH_3CHO + Ag_2O \rightarrow 2Ag^0 + CH_3COOH \tag{16}$$

$$Ag(C_2H_5OH)^+ + e^- \rightarrow Ag^0 + C_2H_5OH \tag{17}$$

The described process is simple, eco-friendly, requires no expensive instruments, and neither surfactants nor additional reducing agents are needed [76].

Recently, environmentally friendly synthesis of AgNP was also achieved by electroreduction of Ag^+ using PVA as an adequate capping agent [79]. The obtained AgNP were spherical, with a mean diameter of ~15 nm as determined by TEM analysis. The use of PVA presents advantages to be considered; low cost, non-toxic, water-soluble, biocompatible and biodegradable. Moreover, the regular linear structure of PVA with a large number of side hydroxyl groups on the main chain suggests excellent hydrophilicity and reactivity. In fact, cyclic voltammetry and FT-IR spectroscopy confirmed that the hydroxyl groups from PVA molecules coordinated with the AgNP, making them more stable during a prolonged period. Thus, AgNP/PVA colloidal dispersions could be utilized for the production of Ag/PVA hydrogels in different forms (thin films, discs and sheets), which could be used for biomedical applications as antimicrobial treatments [79].

Electrochemical methods has been also useful to produce AgNP of different shapes, like nanorods [71], nanoflakes and nanowires [74]. Silver nanorods have been prepared by electroreduction of aqueous solution of $AgNO_3$ in the presence of polyethylene glycol (PEG) as capping material and further characterized by TEM, X-ray and UV-vis spectroscopy [71]. The results shown that both the concentration of $AgNO_3$ and PEG can affect the formation of the nanorods.

Nanowires and nanoflakes were prepared by potentiostatic reduction of Ag^+ on HOPG electrode in acetonitrile solution containing mercaptopropionic acid (MPA) [74]. MPA is immediately adsorbed on the deposited silver and affects further growth of the metal. Due to differences in adsorption of thiol molecules onto the various crystallographic facets of silver, some directions of growth are favored and some are inhibited. In consequence, the deposited silver forms rod and flake-like structures which are distributed on the HOPG surface [74].

In summary, most traditional electrochemical methods for production of AgNP have been proven to present some additional advantages over chemical methods in the synthesis of size-selective or shape-controlled highly pure metal nanomaterials, by simple adjusting the current density, solvent conditions, capping agents, etc.

4.1 Sonoelectrochemistry

Combination of ultrasound irradiation and electrochemistry processes date back to the 1930s but in the last decade the expansion of this methodology has become increasingly important. Sonoelectrochemistry is an alternative method, simple and cost effective to obtain AgNP [81]. The diversity of induced effects on electrochemistry processes by ultrasound waves can be attributed to the generation, growth and collapse of microbubbles within the electrolyte [82].

A number of different instrumental setups have been used for the incorporation of the ultrasound irradiation into the electrochemical systems. The first and simplest setup used was the immersion of a conventional electrochemistry cell at a fixed position in an ultrasounic bath. One of the most adequate systems was first described by Reisse et al. [83] (Fig. 16). This system used the simplest configuration of a

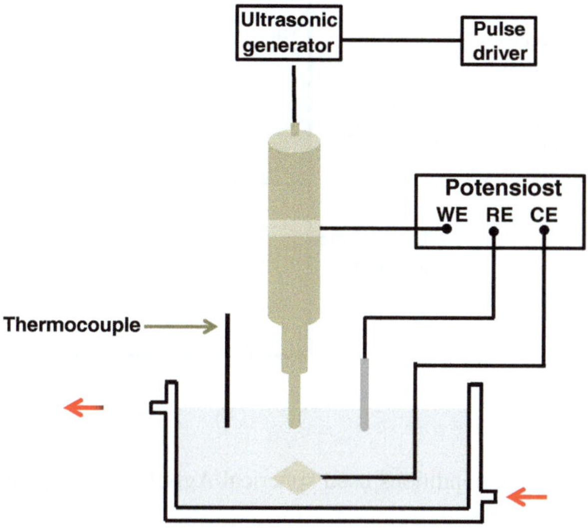

Fig. 16 Sonoelectrochemistry set-up used in the production of nanopowders (*WE* working, *RE* reference and *CE* auxiliary electrode) according to Ref. [81]

two-electrode cell because the process is under galvanostatic conditions. However, undesirable secondary reactions were observed, and so an adaptation was made. The replacement of the two-electrode configuration (cathode and anode) by a three-electrode configuration (working, reference and auxiliary electrodes) in the sonoelectrochemistry system was performed with the aim of applying a controlled potential to the sonoelectrode to obtain better control over the process.

The fundamental basis of the pulsed sonoelectrochemical technique for the production of nanopowders is massive nucleation. At the cathode, a pulse of current (or potential) reduces a number of cations, depositing a high density of metal nuclei on the sonoelectrode surface, and the titanium horn works only as an electrode during this time (T_{ON}). After this short electrochemical pulse, a short ultrasound pulse of high intensity (T_{US}) removes the metal particles from the cathode surface and replenishes the double layer with metal cations by stirring the solution. Sometimes, a rest time (T_{OFF}) without current or ultrasonic vibrations follows the two previous pulses and it is useful to restore the initial conditions close to the sonoelectrode surface. Figure 17 shows the pulse distribution with time. Electrochemical and ultrasound pulses typically ranges between 100 and 500 ms and the rest time lasts no more than 1 s.

Silver nanoparticles have been synthesized using different electrolytes and stabilizers by the pulsed sonoelectrochemistry method. AgNP with different morphology including spheres, rods and dendrites were prepared from an aqueous solution of $AgNO_3$ in the presence of nitriloacetate (NTA) [84].

Jiang et al. [85] reported the synthesis of silver nanoparticles with a face-centered cubic (fcc) structure in a saturated solution of silver citrate in the presence of PVP.

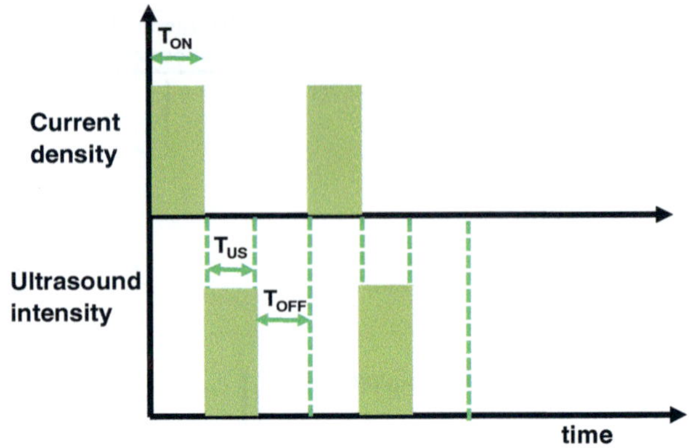

Fig. 17 Representation of ultrasound and current pulse distribution with time according to Ref. [69]. (T_{ON} current pulse time, T_{US} ultrasound pulse time and T_{OFF} rest time)

Under the experimental conditions used spherical AgNP were prepared with an average diameter of 20–25 nm.

Some of the advantages of this method are: acceleration of mass transport, cleaning and degassing of the electrode surface, and an increased reaction rate [86].

Process yield and particle size are affected by various factors such as the ultrasound pulse time and the current density. In general, decreasing temperature, shorten pulse duration, high current density and high ultrasound intensity will lead to a reduction in crystal size. These parameters need to be optimized in order to maximize the nanoparticles production yield and to obtain the smaller size of the products depending of their applications.

In addition, undesirable side effects like aggregation can be reduced by controlling reactions parameters such as electrodeposition and ultrasound conditions, and by using a suitable stabilizer.

5 Characterization Techniques: What Information Do They Give Us?

Characterizing metallic nanoparticles as AgNP implies determination of their size, shape, chemical composition, and for some applications also their bulk concentration. Next, we summarize the most common techniques used to characterize the synthesized metal nanoparticles including AgNP.

Traditionally, chemical characterization of AgNP has been done using transmission or scanning electron microscopy (TEM/SEM) followed by energy-dispersive X-ray spectroscopy (EDS) and X-ray diffraction (XRD). Also, size exclusion and ion chromatography (SEC and IC) with multi-element detection (inductively-coupled plasma mass spectrometry and optical emission spectroscopy, ICP-MS and ICP-OES, respectively) have been applied because of the intrinsic limitations of Electron Microscopy. Other disadvantages of EM are charging effects (caused

Table 1 Common analytical methods for AgNP characterization

Method	Information	Advantages	Limitations
Turbidimetry/Nephelometry	Concentration with standards	Sample preparation	Low sensitivity (mg/L)
		Non invasive and	
		Non destructive	Sample artifacts
Laser-induced breakdown detection (LIBD)	Concentration size	High sensitivity (μg/L)	Dilute solutions
			Research-grade Instrumentation
			External calibrations
			No elemental information
UV-vis spectroscopy	Concentration	Non invasive	Interferences from turbidity
	Some structural information	Non destructive	
Inductively-coupled plasma atomic emission spectroscopy and mass spectrometry ICP-AES/ICP-MS	Elemental composition	Simultaneous elemental composition	Requires acid digestion
		Very sensitive (μg/L)	Destructive
		Selectivity: MS > AES	Expensive instrumentation
			MS requires expensive instrumentation
Near field scanning optical microscopy (NSOM)	Optical image	Non destructive	Low resolution (diffraction limit of light/size of nanoparticles)
	Size and shape (30 nm level)		
Scanning electron microscopy (SEM)	Size and shape (1–1 μm)	High resolution	Destructive
			Charge effects
			Incompatible with wet or liquid samples
Transmission EM (TEM)/Scanning TEM (STEM)	Size	Very high-resolution	Destructive
	Shape		Charge effects
	Structural information (up to 0.1 nm)		Incompatible with wet or liquid samples
			Expensive instrumentation

(continued)

Table 1 (continued)

Method	Information	Advantages	Limitations
Atomic force microscopy (AFM)	Height	Provides 3D surface plots at sub-nm resolution with adequate instrumentation (type of tip, scanner, excellent signal to noise ratio)	Affected by tip coating and loading
	Topography (Resolution < 0.1 nm)		Overestimation of lateral dimensions
Dynamic light scattering (DLS)	Hydrodynamic diameter (>3 nm)	Simple and rapid analysis	Limited capability for polydisperse samples
Electrophoretic mobility (EM)	Zeta potential (ξ)	Simple and rapid analysis	Interpretation in terms of nanoparticle surface

by accumulation of static electric fields), ~10 % of uncertainties in quantitative analysis and to operate under vacuum conditions [87, 88].

Imaging in fluid solutions is possible using atomic force microscopy (AFM), in which 3D surface profiles can be achieved with height resolutions of ~0.5 nm. The main limitation of AFM is that the geometry of the tip is often larger than the particles being probed and this leads to errors in the onset and offset of particle topography on a scan, resulting in overestimations of the lateral dimensions of the nanoparticles [88].

The spectral response of AgNP in the visible and infrared region of the electromagnetic spectrum has attracted even more attention. The phenomenon known as the surface plasmon resonance (SPR) is localized on the surface of the nanostructure [89]. These resonances stem from the collective oscillation of surface conduction electrons driven by the incident electromagnetic field. The SPR properties of metal nanoparticles are strongly size and shape dependent in addition to capping effects and dielectric constant of the medium, see Chap. "Silver Nanoparticles: From Bulk Material to Colloidal Nanoparticles". Thus, control over the size and shape of metal nanoparticles is of considerable importance. In a UV-Vis spectrum, the average particle size is associated with the maximum absorption wavelength, while the particle dispersion is related with the full width at half maximum (fwhm) [87].

Scattering techniques like static (SLS) and dynamic light scattering (DLS), or neutron scattering, such as small-angle neutron scattering (SANS) are also useful for nanoparticle characterization. For example, DLS is particularly employed for sizing nanoparticles and determining their state of aggregation in suspensions [90]. Other laser-based techniques include Raman spectroscopy and laser-induced fluorescence (LIF) allowing a better understanding of physical properties [90].

Thermogravimetry and differential thermo analysis (TG–DTA) are useful for investigating the thermal stability and decomposition, dehydration, oxidation, as well as the determination of volatile content and other compositional analysis. TG in combination with a mass spectrometer can be used for surface analysis [90]. Properties of dispersed particles are also studied by electrophoresis in particular, for measuring the zeta potential that informs about the overall charge of a particle and gives an indication of the stability of a colloidal system against agglomeration [90].

In Table 1 are summarized the main methods available for the analysis of AgNP ordered by the information obtained [87, 88, 90].

6 Concluding Remarks

Silver nanoparticles can be obtained by either a top-down or bottom-up approach. Hundreds of research articles reporting different synthetic methods for AgNP are published every year. Throughout this chapter we have reviewed only some of the most relevant works, dealing mostly with bottom-up: chemical reduction, photochemical and electrochemical methods. The particle formation mechanisms were discussed as a key to understand (and predict) the outcome of any synthetic method.

Depending on the AgNP application, the choice of the synthetic route is not a trivial decision as the product (e.g. size and shape) depends on the metal precursor, capping selection, reaction temperature, etc. Nucleation and growth mechanisms do not always follow classical models (La Mer´s) as it has been described, for example, in the case of chemical reduction using sodium borohydride, and this affects mainly the final size together with polidispersity. For its part, the polyol method represents a chemical alternative to modulate the shape by selecting the adequate PVP/Ag^{+} concentration ratio.

Light-assisted methods for AgNP are attractive due to their high versatility, selectivity and possibility to have control over the final particles shape and size. Electrochemical methods also permit selection of shape and size by adjusting parameters like current density, solvent conditions, pulse duration, etc.

Finally, we expect this chapter to be helpful on deciding the most appropriate method to synthesize AgNP to accomplish the readers' purposes.

Acknowledgments We want to thank to all the researchers whose work has been cited in here. C.D.B thanks the Agencia de Promoción Científica y Tecnológica (ANPCyT) and Universidad Nacional de Santiago del Estero (UNSE) for combined financial support (PICTO-UNSE-2012-0013). N.L.P and A.V.V would like to thank funding support from CONICET, ANPCyT-PICT 2011-0106, SECYT-UNC and Mincyt-Córdoba. N.P., A.V., V.R., and C.D.B. are research members of the Consejo Nacional de Investigaciones Científicas y Técnicas (CONICET) of Argentina.

References

1. Rogers, K.R., et al.: Alterations in physical state of silver nanoparticles exposed to synthetic human stomach fluid. Sci. Total Environ. **420**, 334–339 (2012)
2. Alarcon, E.I., et al.: The biocompatibility and antibacterial properties of collagen-stabilized, photochemically prepared silver nanoparticles. Biomaterials **33**(19), 4947–4956 (2012)
3. Lee, P.C., Meisel, D.: Adsorption and surface-enhanced raman of dyes on silver and gold sol. J. Phys. Chem. **86**, 3391–3395 (1982)
4. Li, W., et al.: Dimers of silver nanospheres: facile synthesis and their use as hot spots for surface-enhanced raman scattering. Nano Lett. **9**, 485–490 (2009)
5. Alvarez-Puebla, R.A., Aroca, R.F.: Synthesis of silver nanoparticles with controllable surface charge and their application to surface-enhanced raman scattering. Anal. Chem. **81**, 2280–2285 (2009)
6. Stamplecoskie, K.G., Scaiano, J.: Optimal size of silver nanoparticles for surface-enhanced raman spectroscopy. J. Phys. Chem. C **115**, 1403–1409 (2011)
7. Marsich, L., et al.: Poly-l-lysine-coated silver nanoparticles as positively charged substrates for surface-enhanced raman scattering. Langmuir **28**, 13166–13171 (2012)
8. Li, J.M., et al.: Detecting trace melamine in solution by SERS using Ag nanoparticle coated poly(styrene-co-acrylic acid) nanospheres as novel active substrates. Langmuir **27**(23), 14539–14544 (2011)
9. Wang, B., Zhang, L., Zhou, X.: Synthesis of silver nanocubes as a SERS substrate for the determination of pesticide paraoxon and thiram. Spectrochim. Acta Part A Mol. Biomol. Spectrosc. **121**, 63–69 (2014)
10. Hornyak, G.L., et al.: Introduction to Nanosciences. CRC Press. Taylor & Francis Group, Boca Raton (2008)

11. Narayanan, K.B., Sakthivel, N.: Biological synthesis of metal nanoparticles by microbes. Adv. Colloid Interface Sci. **156**(1–2), 1–13 (2010)
12. Hebbalalu, D., et al.: Greener techniques for the synthesis of silver nanoparticles using plant extracts, enzymes, bacteria, biodegradable polymers, and microwaves. ACS Sustain. Chem. Eng. **1**(7), 703–712 (2013)
13. Rycenga, M., et al.: Controlling the synthesis and assembly of silver nanostructures for plasmonic applications. Chem. Rev. **111**(6), 3669–3712 (2011)
14. Sun, Y.: Controlled synthesis of colloidal silver nanoparticles in organic solutions: empirical rules for nucleation engineering. Chem. Soc. Rev. **42**, 2497–2511 (2013)
15. Sakamoto, M., Fujistuka, M., Majima, T.: Light as a construction tool of metal nanoparticles: synthesis and mechanism. J. Photochem. Photobiol. C **10**(1), 33–56 (2009)
16. Cushing, B.L., Kolesnichenko, V.L., Oconnor, C.J.: Recent advances in the liquid-phase syntheses of inorganic nanoparticles. Chem. Rev. **104**, 3893–3946 (2004)
17. Vanysek, P.: Electrochemical series. In: Lide, D.R. (ed) CRC Handbook of Chemistry and Physics, p. 8.21–8.31. CRC Press, LLC (2003–2004)
18. Hoonacker, A.V., Englebienne, P.: Revisiting silver nanoparticle chemical synthesis and stability by optical spectroscopy. Curr. Nanosci. **2**, 359–371 (2006)
19. Hudnall, P.M.: Hydroquinone. In: Ullmann's Encyclopedia of Industrial Chemistry. Wiley-VCH Verlag GmbH & Co, KGaA (2000)
20. Turkevich, J., Stevenson, P.C., Hillier, J.: A study of the nucleation and growth processes in the synthesis of colloidal gold. Discuss. Faraday Soc. 55–75, (1951)
21. Pillai, Z.S., Kamat, P.V.: What factors control the size and shape of silver nanoparticles in the citrate ion reduction method? J. Phys. Chem. B **108**, 945–951 (2004)
22. Henglein, A., Giersig, M.: Formation of colloidal silver nanoparticles: capping action of citrate. J. Phys. Chem. B **103**, 9533–9539 (1999)
23. Segur, J.B., Oberstar, H.E.: Viscosity of glycerol and its aqueous solutions. Ind. Eng. Chem. **43**(9), 2117–2120 (1951)
24. Steinigeweg, D., Schlücker, S.: Monodispersity and size control in the synthesis of 20–100 nm quasi-spherical silver nanoparticles by citrate and ascorbic acid reduction in glycerol–water mixtures. Chem. Commun. **48**(69), 8682–8684 (2012)
25. Caswell, K.K., Bender, C.M., Murphy, C.J.: Seedless, surfactantless wet chemical synthesis of silver nanowires. Nano Lett. **3**(5), 667–669 (2003)
26. Van Hyning, D.L., Zukoski, C.F.: Formation mechanisms and aggregation behavior of borohydride reduced silver particles. Langmuir **14**, 7034–7040 (1998)
27. Viswanatha, R., Sarma, D.: Growth of nanocrystals in solution. In: Rao, C.N.R., Müller, A. Cheetham A.K. (eds.) Nanomaterials Chemistry, p. 139–170. WILEY-VCH, Weinheim (2007)
28. La Mer, V.K., Dinegar, R.H.: Theory, production and mechanism of formation of monodisperse hydrosols. J. Am. Chem. Soc. **72**, 4847–4854 (1950)
29. Polte, J., et al.: Formation mechanism of colloidal silver nanoparticles: analogies and differences to the growth of gold nanoparticles. ACS Nano **6**(7), 5791–5802 (2012)
30. Wuithschick, M., et al.: Size-controlled synthesis of colloidal silver nanoparticles based on mechanistic understanding. Chem. Mater. **25**, 4679–4689 (2013)
31. Perez, M.A., et al.: Hydroquinone synthesis of silver nanoparticles: a simple model reaction to understand the factors that determine their nucleation and growth. Cryst. Growth Des. **8**, 1377–1383 (2008)
32. Patakfalvi, R., Dekany, I.: Nucleation and growth of silver nanoparticles monitored by titration microcalorimetry. J. Therm. Anal. Calorim. **79**, 587–594 (2005)
33. Yoosaf, K., et al.: In situ synthesis of metal nanoparticles and selective naked-eye detection of lead ions from aqueous media. J. Phys. Chem. C **111**, 12839–12847 (2007)
34. Gallardo, O., et al.: Silver oxide particles/silver nanoparticles interconversion: susceptibility of forward/backward reactions to the chemical environment at room temperature. RSC Adv. **2**(7), 2923 (2012)

35. Wan, Y., et al.: Quasi-spherical silver nanoparticles: Aqueous synthesis and size control by the seed-mediated Lee-Meisel method. J. Colloid Interface Sci. **394**, 263–268 (2013)
36. Wiley, B., Sun, Y., Xia, Y.: Synthesis of silver nanostructures with controlled shapes and properties. Acc. Chem. Res. **40**(10), 1067–1076 (2007)
37. Burda, C., et al.: Chemistry and properties of nanocrystals of different shapes. Chem. Rev. **105**(4), 1025–1102 (2005)
38. Pastoriza-Santos, I., Liz-Marzán, L.M.: Formation and stabilization of silver nanoparticles through reduction by N,N-Dimethylformamide. Langmuir **15**, 948–951 (1999)
39. Pastoriza-Santos, I., Liz-Marzán, L.M.: Synthesis of silver nanoprisms in DMF. Nano Lett. **2**(8), 903–905 (2002)
40. Rodríguez-Gattorno, G., et al.: Metallic nanoparticles from spontaneous reduction of silver(I) in DMSO. Interaction between nitric oxide and silver nanoparticles. J. Phys. Chem. B **106**, 2482–2487 (2002)
41. Sun, Y., Xia, Y.: Shape-controlled synthesis of gold and silver nanoparticles. Science **298**(5601), 2176–2179 (2002)
42. Sun, Y., et al.: Polyol synthesis of uniform silver nanowires: a plausible growth mechanism and the supporting evidence. Nano Lett. **3**(7), 955–960 (2003)
43. Sun, Y., et al.: Uniform silver nanowires synthesis by reducing AgNO3 with ethylene glycol in the presence of seeds and poly(Vinyl Pyrrolidone). Chem. Mater. **14**, 4736–4745 (2002)
44. Wiley, B., et al.: Polyol synthesis of silver nanoparticles: use of chloride and oxygen to promote the formation of single-crystal, truncated cubes and tetrahedrons. Nano Lett. **4**(9), 1733–1739 (2004)
45. Lin, J.-Y., Hsueh, Y.-L., Huang, J.-J.: The concentration effect of capping agent for synthesis of silver nanowire by using the polyol method. J. Solid State Chem. (2014)
46. Tao, A., Habas, S., Yang, P.: Shape control of colloidal metal nanocrystals. Small **4**(3), 310–325 (2008)
47. Wiley, B., et al.: Shape-controlled synthesis of metal nanostructures: the case of silver. Chem. Eur. J. **11**(2), 454–463 (2005)
48. Stamplecoskie, K., Scaiano, J.: Silver as an example of the applications of photochemistry to the synthesis and uses of nanomaterials. Photochem. Photobiol. **88**(4), 762–768 (2012)
49. Mafuné, F., et al.: Structure and stability of silver nanoparticles in aqueous solution produced by laser ablation. J. Phys. Chem. B **104**(35), 8333–8337 (2000)
50. Bae, C.H., Nam, S.H., Park, S.M.: Formation of silver nanoparticles by laser ablation of a silver target in NaCl solution. Appl. Surf. Sci. **197–198**, 628–634 (2002)
51. Tsuji, T., Okazaki, Y., Tsuji, M.: Photo-induced morphological conversions of silver nanoparticles prepared using laser ablation in water—Enhanced morphological conversions using halogen etching. J. Photochem. Photobiol. A **194**(2–3), 247–253 (2008)
52. Jiménez, E., et al.: A novel method of nanocrystal fabrication based on laser ablation in liquid environment. Superlattices Microstruct. **43**(5–6), 487–493 (2008)
53. Hada, H., et al.: Photoreduction of silver ion in aqueous and alcoholic solutions. J. Phys. Chem. **80**(25), 2728–2731 (1976)
54. Guang-Nian, X., et al.: Preparation and characterization of stable monodisperse silver nanoparticles via photoreduction. Colloids Surf. A **320**(1–3), 222–226 (2008)
55. Huang, H.H., et al.: Photochemical formation of silver nanoparticles in poly(N-vinylpyrrolidone). Langmuir **12**(4), 909–912 (1996)
56. Gaddy, G.A., et al.: Photogeneration of silver particles in PVA fibers and films. J. Clust. Sci. **12**(3), 457–471 (2001)
57. Huang, H.T., Yang, Y.: Preparation of silver nanoparticles in inorganic clay suspensions. Compos. Sci. Technol. **68**(14), 2948–2953 (2008)
58. Chegel, V., et al.: Ag nanoparticle-poly(acrylic acid) composite film with dynamic plasmonic properties. Aust. J. Chem. **65**(9), 1223–1227 (2012)
59. Scaiano, J.C., et al.: Photochemical routes to silver and gold nanoparticles. Pure Appl. Chem. **81**(4), 635–647 (2009)

60. Scaiano, J.C., et al.: Magnetic field control of photoinduced silver nanoparticle formation. J. Phys. Chem. B **110**(26), 12856–12859 (2006)
61. Alarcon, E., et al.: Human serum albumin as protecting agent of silver nanoparticles: role of the protein conformation and amine groups in the nanoparticle stabilization. J. Nanopart. Res. **15**(1), 1374 (2013)
62. Jockusch, S., et al.: Photochemistry and photophysics of α-Hydroxy ketones. Macromolecules **34**(6), 1619–1626 (2001)
63. Gonzalez, C.M., Liu, Y., Scaiano, J.C.: Photochemical strategies for the facile synthesis of gold-silver alloy and core-shell bimetallic nanoparticles. J. Phys. Chem. C **113**(27), 11861–11867 (2009)
64. McGilvray, K.L., et al.: Photochemical strategies for the seed-mediated growth of gold and gold—silver nanoparticles. Langmuir **28**(46), 16148–16155 (2012)
65. Maretti, L., et al.: Facile photochemical synthesis and characterization of highly fluorescent silver nanoparticles. J. Am. Chem. Soc. **131**(39), 13972–13980 (2009)
66. Stamplecoskie, K.G., Scaiano, J.: Kinetics of the formation of silver dimers: early stages in the formation of silver nanoparticles. J. Am. Chem. Soc. **133**(11), 3913–3920 (2011)
67. Callegari, A., Tonti, D., Chergui, M.: Photochemically grown silver nanoparticles with wavelength-controlled size and shape. Nano Lett. **3**(11), 1565–1568 (2003)
68. Stamplecoskie, K.G., Scaiano, J.: Light emitting diode irradiation can control the morphology and optical properties of silver nanoparticles. J. Am. Chem. Soc. **132**(6), 1825–1827 (2010)
69. Rodríguez-Sánchez, L., Blanco, M.C., Lopez-Quintela, M.: Electrochemical synthesis of silver nanoparticles. J. Phys. Chem. B **104**(41), 9683–9688 (2000)
70. Tang, Z., et al.: Electrochemical synthesis of Ag nanoparticles on functional carbon surfaces. J. Electroanal. Chem. **502**(1–2), 146–151 (2001)
71. Zhu, J.-J., et al.: Preparation of silver nanorods by electrochemical methods. Mater. Lett. **49**(2), 91–95 (2001)
72. Ueda, M., et al.: Double-pulse technique as an electrochemical tool for controlling the preparation of metallic nanoparticles. Electrochim. Acta **48**(4), 377–386 (2002)
73. Ma, H., et al.: Synthesis of silver and gold nanoparticles by a novel electrochemical method. Chem. Phys. Chem. **5**(1), 68–75 (2004)
74. Mazur, M.: Electrochemically prepared silver nanoflakes and nanowires. Electrochemi. Commun. **6**(4), 400–403 (2004)
75. Jian, Z., Xiang, Z., Yongchang, W.: Electrochemical synthesis and fluorescence spectrum properties of silver nanospheres. Microelectron. Eng. **77**(1), 58–62 (2005)
76. Starowicz, M., Stypuła, B., Banaś, J.: Electrochemical synthesis of silver nanoparticles. Electrochem. Commun. **8**(2), 227–230 (2006)
77. Hu, M.Z., Easterly, C.E.: A novel thermal electrochemical synthesis method for production of stable colloids of "naked" metal (Ag) nanocrystals. Mater. Sci. Eng. C **29**(3), 726–736 (2009)
78. Khaydarov, R., et al.: Electrochemical method for the synthesis of silver nanoparticles. J. Nanopartic. Res. **11**(5), 1193–1200 (2009)
79. Surudzic, R., et al.: Electrochemical synthesis of silver nanoparticles in poly(vinyl alcohol) solution. J. Serb. Chem. Soc. **78**(12), 2087–2098 (2013)
80. Plieth, W., et al.: Electrochemical preparation of silver and gold nanoparticles: characterization by confocal and surface enhanced Raman microscopy. Surf. Sci. **597**(1–3), 119–126 (2005)
81. Sáez, V., Mason, T.: Sonoelectrochemical synthesis of nanoparticles. Molecules **14**(10), 4284–4299 (2009)
82. Compton, R.G., Eklund, J.C., Marken, F.: Sonoelectrochemical processes: a review. Electroanal **9**(7), 509–522 (1997)
83. Reisse, J., et al.: Sonoelectrochemistry in aqueous electrolyte: a new type of sonoelectroreactor. Electrochim. Acta **39**(1), 37–39 (1994)

84. Zhu, J., et al.: Shape-controlled synthesis of silver nanoparticles by pulse sonoelectrochemical methods. Langmuir **16**, 6396–6399 (2000)
85. Jiang, L.-P., et al.: A novel route for the preparation of monodisperse silver nanoparticles via a pulsed sonoelectrochemical technique. Inorg. Chem. Commun. **7**(4), 506–509 (2004)
86. Mason, T.J., Lorimer, J.P., Walton, D.J.: Sonoelectrochemistry. Ultrasonics **28**(5), 333–337 (1990)
87. Liu, J., et al.: Methods for separation, identification, characterization and quantification of silver nanoparticles. TrAC Trends Anal. Chem. **33**, 95–106 (2012)
88. Ferreira da Silva, B., et al.: Analytical chemistry of metallic nanoparticles in natural environments. TrAC, Trends Anal. Chem. **30**(3), 528–540 (2011)
89. Zheng, X., et al.: Photochemical formation of silver nanodecahedra: structural selection by the excitation wavelength. Langmuir **25**, 3802–3807 (2009)
90. Tiede, K., et al.: Detection and characterization of engineered nanoparticles in food and the environment. Food Addit. Contam. **25**, 795–821 (2008)

Surface Enhanced Raman Scattering (SERS) Using Nanoparticles

Altaf Khetani, Ali Momenpour, Vidhu S. Tiwari and Hanan Anis

Abstract Surface enhanced Raman spectroscopy (SERS) is one of the most promising label-free techniques for molecular sensing. In this chapter, we review the fundamental concepts of Raman spectroscopy and the need for SERS. We then discuss the fundamentals of SERS and focus on silver nanoparticles. This is followed by reviewing recent progress and applications of SERS. Finally, we discuss methods for further enhancing SERS signal using hollow core fibers.

1 Introduction

Raman spectroscopy is an effective analytical technique used to obtain optical 'fingerprints' of molecules. Raman spectra usually contain many sharp peaks that correspond to specific molecular vibrational frequencies, and these can provide a clear signature defining the presence of specific molecules in a sample. Accordingly, Raman spectra can be used to qualitatively and quantitatively discriminate between chemical species in materials [1]. Despite these capabilities, a key limitation of the Raman effect is its extremely weak signal, which means that high power lasers and longer acquisition times are required. This in turn can lead to sample damage and thus limit biological application. Bypassing this shortcoming is achieved by surface enhanced Raman scattering (SERS) [2].

A. Khetani · A. Momenpour · H. Anis (✉)
School of Electrical Engineering and Computer Science,
University of Ottawa, Ottawa, ON K1N 6N5, Canada
e-mail: hanis@uottawa.ca

V.S. Tiwari
SRM University, Sonepat–Kundli Urban Complex Sonepat,
Haryana 131029, India

E.I. Alarcon et al. (eds.), *Silver Nanoparticle Applications*, Engineering Materials,
DOI 10.1007/978-3-319-11262-6_3

SERS, used for extremely low concentrations of molecules overcomes the low Raman cross-section barrier by exploiting the large field enhancement caused by electromagnetic coupling between the nanoparticles. Under such conditions, the Raman signal of the target molecule is enhanced by several orders of magnitude, enabling detection down to a single molecule [3–6]. In general, SERS is performed on a gold (Au), silver (Ag) or copper (Cu) colloidal or substrate. The majority of SERS experiments utilizes Au and Ag, as they are air stable and the absorbance is mostly in visible and near IR wavelengths. Over the years, researchers have strived to optimize the structure, shape, volume-ratio and configuration to obtain maximum enhancement of SERS signals.

For highly optimized configurations, enhancement of the Raman signal in the order of 10^6–10^8 has been obtained. There are basically two main reasons for the amplified Raman signal. One is the strong enhancement of the electromagnetic field by the localized surface plasmon resonance (SPR) of metallic nanostructures, which is considered the main contributor to the SERS enhancement factor. The other reason is the surface chemical enhancement mechanism [7, 8].

SERS has been predominantly used as an analytical technique in the fields of biochemistry, forensics, food safety, threat detection and medical diagnostics [4, 9–11]. The reasons for the popularity of SERS are that it offers: (i) molecular fingerprints down to the level of single molecule detection, (ii) label-free detection as opposed to fluorescence spectroscopy, (iii) multiplex detection, (iv) minimal sample requirements, and (v) in situ and in vitro characterization of biological samples.

'Multivariate analysis' has played an important role in SERS spectroscopy as it is used for quantitative analysis, particularly when there is an overlap of the Raman bands of interest with those of the sampling media (e.g. serum, blood). It provides a simple calibration curve between the Raman signal intensity and the sample concentration. Multivariate methods, such as Partial Least Squares (PLS) perform the quantitative spectral data analysis by constructing models in which the response Y-variable (analytical data) depends on more than one explanatory X-variable (Raman shift wave number). During the formulation of the calibration model, linear combinations of only X-variables that are related to the response 'Y-variable' are considered. Thus, PLS has been successfully applied to quantitatively and qualitatively identify components in complex mixtures, tissues and cells [12].

The increasing interest in SERS can be seen by the growth in the number of publications shown in Fig. 1, where SERS had been used in the past two to three decades. Although different metallic nanoparticles have been used in SERS studies, this chapter mainly focuses on SERS and its applications using silver nanoparticles (AgNP). Section 2 in this chapter revises the theory and principles behind SERS, and in Sect. 3 we outline the use of AgNP in SERS applications. In Sects. 4 and 5, we summarize recent progress and applications in the SERS field. Finally, in Sect. 6, recent advances in the use of AgNP for SERS mediated analysis of biomolecules (rhodamine, adenosine) with novel photonic crystal fiber is presented.

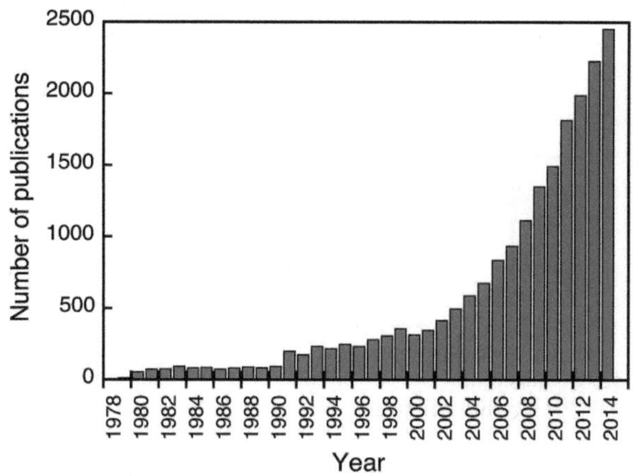

Fig. 1 Representation of the number of research articles published in the period 1979–2014 showing the popularity of the surface enhanced Raman technique. Data obtained from Web of Science search using the term "surface enhanced Raman" accessed on May 26th, 2014

2 Surface Enhanced Raman Scattering (SERS)

2.1 Fundamentals of SERS

Raman scattering is an inelastic phenomenon that occurs when a sample is illuminated by a laser light, as first demonstrated by Raman and Krishnan in 1928 [13]. Most of the photons emitted after scattering have the same frequency as the incident photons (Rayleigh scattering) while a relatively small number of the scattered photons (approximately 1 in 10^6–10^{10} photons [14, 15]) have shifted frequencies. The downshift or upshift in frequency is designated as stokes or anti-stokes Raman scattering, and provides information about vibrational transition in molecules. Since only a few photons are Raman scattered, there has been considerable interest in how to increase the weak Raman signal.

SERS is a method widely used to enhance the intensity of the Raman signal. First observed by Fleischmann in 1974 [16], SERS relies on adsorption of the analyte onto the surface of metal structures (typically silver, gold or copper). Raman intensity of molecules adsorbed on the surface of specific metals can be 10^6–10^8 times stronger than the normal Raman intensity of that molecule. This illustrates the importance of SERS and its association with metal structures, the subject matter of this chapter and discussed in the following sections.

Raman is proportional to the square of the electric dipole moment of the molecule ($p = \alpha E$). The polarizability (α) or electric fields (E) are two possible reasons for this enhancement. The enhancement due to polarizability is called chemical enhancement, and that due to electric fields is called electromagnetic (EM) mechanism. In EM enhancement, the mechanism of SERS is described by the electromagnetic model, which considers a small amount of illuminated metal in an electromagnetic field. According to the EM mechanism, the electric field near a metal particle is enhanced because of surface plasmon excitation. The surface plasmon is a confined excitation of electron gas near the surface of the metal. This enhancement in the local electric field causes more intense electronic transition in

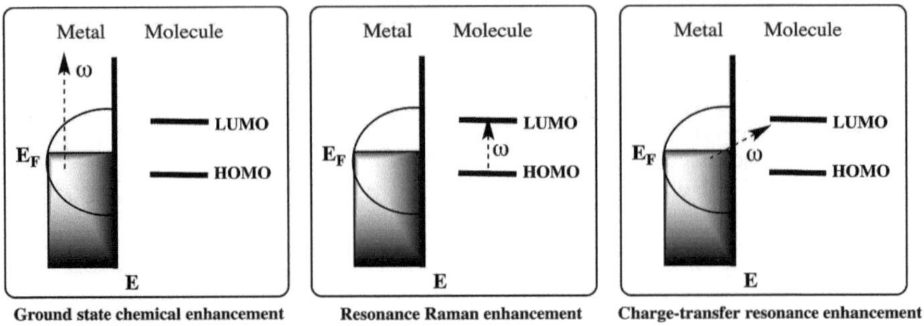

Fig. 2 Illustration of the different types of enhancement mechanism in SERS. Adapted from Ref. [19]. Royal Society of Chemistry® 2008

the analyte adsorbed at the metal surface, and leads to an enhanced Raman signal [17]. In this model, it is assumed that the metal diameter is much smaller than the wavelength of the exciting light.

The other enhancement mechanism via a chemical route can manifest in three different ways [18, 19]. Figure 2 shows the different types of enhancement mechanisms, with the highest occupied molecular orbit (HOMO) and the lowest unoccupied molecular orbit (LUMO) indicated. The first chemical mechanism, known as *ground state chemical enhancement*, is the simplest case. It can occur when the adsorbate does not bind covalently to the metal. In this case, enhancement takes place due to ground state chemical interactions between the molecule and nanoparticles that are not associated with any excitations of the nanoparticle-molecule system. The presence of the metal perturbs the electronic structure of the analyte, causing a 'mild' change in its electronic distribution. In this case, the charge transfer (CT) is not required.

The Second, *resonance Raman enhancement*, involves the presence of the nanoparticle surface-molecule complex, either by direct (covalent) binding to the metal or by indirect binding with the help of an electrolyte ion (typically chloride). This surface complex changes the intrinsic polarizability of the molecule. The surface complex can also create a new electronic state that is explicitly in resonance, or close to being in resonance with the laser, thus contributing to the enhancement of a resonant Raman type.

Lastly, *charge-transfer resonance*, the last of the chemical phenomena, and this is a more sophisticated version of resonance Raman enhancement, involving charge transfer between the analyte and the metal. This can occur when the difference between the Fermi level (E_F) of the metal and the HOMO or LUMO energies of the molecule is matched by the laser, and the excitation wavelength resonates with the nanoparticle–molecule charge transfer transitions.

Overall, the relative contribution of the different mechanisms depends on the experimental conditions but understanding the chemical enhancement contributions to SERS is complex, and requires very accurate electronic structure calculations of the molecule-metal system.

The contribution of the chemical mechanism to the overall enhancement of SERS is much less than that of the EM mechanism. The enhancement factor (EF), a parameter used to evaluate the SERS, shows that SERS that is due to the EM mechanism, is approximately 10^6, as compared to it is 10–100 for the chemical mechanism. The overall enhancement of SERS, by the superposition of EM enhancement and CT enhancement, can thus be in the 10^6–10^8 range [20]. Essentially, it has been shown that the use of stronger electromagnetic fields leads to higher enhancement factors.

2.2 *Localized Surface Plasmon Resonance*

As discussed in the previous section, enhancement resultant from the EM is due to surface plasmon resonance or SPR that is produced near the nanoparticle and the studied molecule. Free electron charges on the metallic nanoparticles respond to external electromagnetic fields (laser light), and oscillate at resonance wavelengths.

Localized surface plasmon resonance (LSPR) depends strongly on the optical properties of metallic nanoparticles, which is described by a complex dielectric constant (ϵ) or a complex refractive index ($m = \sqrt{\epsilon}$):

$$\in (\lambda) \ = \in_r (\lambda) + i \in_i (\lambda) \tag{1}$$

$$m = n + ik \tag{2}$$

where n is the refractive index, and k is the absorption coefficient of nanoparticles. The LSPR depends on the wavelength of incident light due to the wavelength-dependency of the dielectric constant of nanoparticles. For a metallic sphere in the presence of an external field show that LSPR can be observed when the nanoparticle diameter, d, is far smaller than the laser light ($d \ll \lambda$). In this circumstance, the electric field of the light can be considered uniform, and electrostatic equations can be solved [21]. The solution of Maxwell's equations for a spheroid metallic particle, leads to an expression for the extinction $E(\lambda)$ (the sum of absorption and scattering) of a nanoparticle:

$$E(\lambda) \propto \frac{\in_i (\lambda)}{(\in_r (\lambda) + \chi \in_{med})^2 + \in_i (\lambda)^2} \tag{3}$$

This relation shows that the electric field depends on the dielectric constants of nanoparticles. The different metals (silver, gold and copper) and their diverse optical constants represent different enhancement. For example, AgNP provide 10–100 times higher efficiency than gold nanoparticles [21]. The other parameter, known as the shape factor (χ), describes the deviation from spherical particle

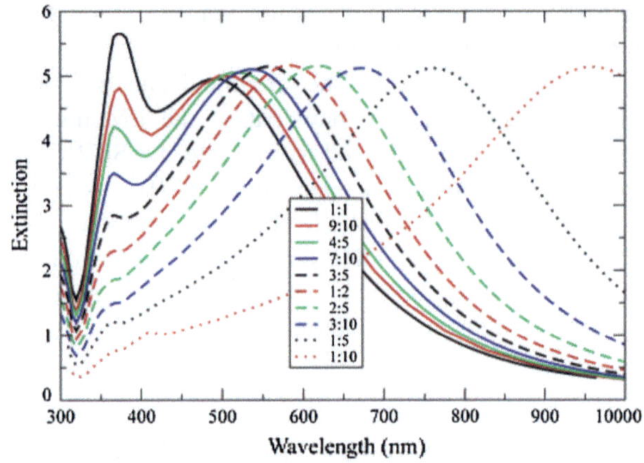

Fig. 3 The extinction spectra of different spheroids, with the same equivalent volume, corresponding to a sphere radius of 80 nm. Reprinted with permission from Ref. [22]. Copyright (2002) American Chemical Society

geometrics into higher aspect ratio structures. This relation shows that the extinction of nanoparticles is strongly dependent on particle shape, and Fig. 3 depicts the dependency of plasmon resonance on particle shape. The major to minor axis ratio, r, varies from 10 to 1. Figure 3 shows the red shift in the peak as the particle becomes more oblate [22].

The red shift also appears when the nanoparticle size is increased. SERS increases with particle size because the enhancement of electromagnetic field intensity depends upon the number of atoms that are excited and the volume of the nanostructure [3]. This effect on surface plasmon is not limited to spherical particles. To illustrate this, the SERS of silver nanorods with an aspect ratio of 10, is 10–100 times higher than that of a sphere, and this is simply due to shape [23]. For nanoparticles with sharper corners and edges producing higher SERS enhancement [24] and metal with a rough surface causes more SERS, due to more localized surface plasmon and stronger field gradients [10]. Specific roughness causes some activity in the surface plasmon, and changes the resonant frequency which enables more scattering [25].

The laser wavelength is more critical when the surface plasmon absorption spectrum is relatively narrow. In this case, the wavelength dependence is very critical, because for SERS the surface plasmon is excited by laser light. The broadening of surface plasmon absorption, which depends on the aggregation where higher aggregation means broader plasmon that is accompanied by a hypsochromic shift [22]. In other words, if the plasmon peak is at λ_1, when aggregating occurs it shifts to λ_2, where $\lambda_2 > \lambda_1$. Making the nanoparticles less stable and forcing them to aggregate creates a collection of different absorption bands. However, enhancement depends on the degree of overlap of the excitation wavelength with the plasmon frequency, which as discussed above is shifted during nanoparticle aggregation. Further, calculations show that the EM enhancement is strongly (i.e. the inverse twelfth power) dependent on the metal-molecule distance. With increasing distance, the EF decreases due to the declining intensity of the dipole moment. Notably, however, this does not mean that enhancement requires direct contact of the metal surface and the molecule [26].

3 Recent Progress in SERS

SERS can provide orders of magnitude increases in Raman intensity, thereby overcoming the traditional drawback of Raman scattering: its inherent weakness. Over the last decade, various approaches to enhance the Raman signal have been attempted, with a wide range of analytes.

Recently, a great deal of attention has been paid to sensitivity. Some researchers have applied SERS for the detection of biologically relevant small molecules. Rapid and accurate detection of bioagents was the aim of Van Duyne's group, who used SERS to detect anthrax biomarkers [27]. They described a procedure for the rapid extraction of CaDPA from B. subtilis spores, simulants for B. anthracis spores, followed by SERS detection on reproducible and stable silver film over a nanosphere substrate. They used silver film over nanosphere (AgFON) as a SERS substrate, and measured the spore concentration range down to 10^{-14} M. A group from the University of Georgia at Athens (UGA) placed rows of silver nanorods on a slide, and detected biological agents or pathogen at attomolar (10^{-18} M) levels [28]. Other research related to highly sensitive SERS detection was performed by Hongyan Liang et al., who prepared monodispersed "flower-like" AgNP [29]. These nanoparticles have a rough surface and they were used for malachite green isothiocyanate (MGITC) molecules detection, at concentrations down to 10^{-10} molar. By comparing the intensity of molecular peaks, they demonstrated that the sensitivity of this type of nanoparticles is 10^6–10^8 times higher than that of normal Raman scattering.

One of the main challenges of SERS is long-term stability and reproducibility. The well-ordered metal nanostructure approach is one of the promising ways to create a stable and reproducible spectrum. In this SERS substrate, the periodic nanostructure of the substrate is covered by the exciting laser. Because the spot of the exciting beam is usually in the range of microns, the entire SERS substrate is excited homogenously. Baia et al., reported the use of a corrugated gold film onto highly ordered polystyrene nanospheres, as SERS-active substrates. SERS experiments for the detection of R6G were efficient to detect discrete molecules adsorbed onto the surface [30]. Zhang et al. investigated and developed a SERS substrate by physical vapour deposition of silver nanolayers onto different types of paper. They demonstrated the detection of analyte concentration down to 10^{-10} M [31], which presents an important advance toward the development of a low cost SERS sensor.

4 SERS: Applications

In this section, we revise four relevant SERS applications. The first is based on AgNP for DNA and tumorigenic cells detection. The second deals detection of toxic substances and the third is the application of SERS for forensic applications. Lastly, we discuss the value of SERS in single molecule detection.

4.1 SERS for Biomedical Sensing Applications

SERS has become a widely used technique in the detection of biological substances including biomolecules (DNA and proteins) and cancer cells [32–42]. AgNP based SERS has been used to detect nasopharyngeal cancer [34], gastric cancer [35, 36], hepatocellular carcinoma [37], esophageal cancer [35], oral cancer [39] and breast cancer [40]. In some of these studies, AgNP was mixed with whole blood to detect circulating cancer cells. Feng et al. developed simple blood plasma biochemical analysis for non-invasive nasopharyngeal cancer detection [34]. Statistical techniques, such as principal component analysis (PCA) combined with linear discriminate analysis (LDA), were applied to the SERS spectra to test the accuracy of the results in differentiating cancer from normal tissues; see Fig. 4 [34]. In another study, Lin et al. used AgNP based SERS technology to analyze and classify human blood plasma, in order to develop a simple and label-free blood test for esophageal cancer detection [35].

SERS based DNA sensors are gaining importance, as the existing fluorescence methods can be affected by dye photobleaching, and other interferences like background scattering. Thus, due to its unique characteristics, including, narrow spectral band (<1 nm, half-widths), and reduced susceptibility to both photobleaching and self-quenching of the dye, AgNP based SERS has also been explored for DNA sensing (see [41–46]).

Figure 5 shows the color change in aggregated NPs when mixed with DNA bases, as well their spectra [43]. Wang et al., demonstrated a proof-of-principle concept, in which an AgNP based SERS nanoprobe was used for early microbial detection in patients [47]. As a model for the approach, the nanoprobe was used

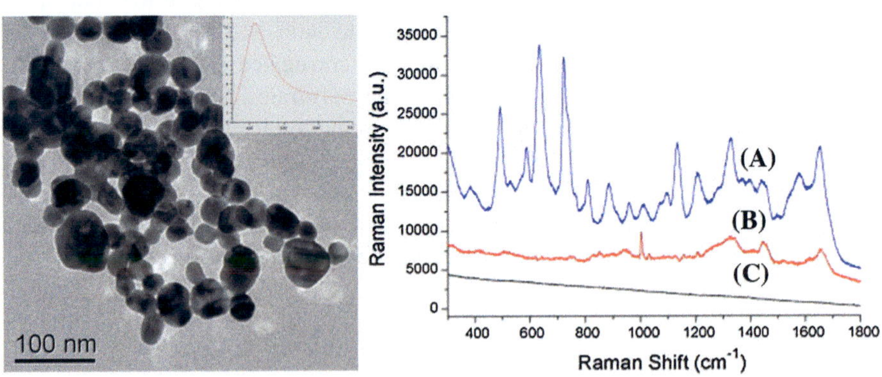

Fig. 4 **a** TEM micrograph of Ag colloidal surface with UV/visible absorption curve of the pure Ag colloid. **b** (*A*) SERS spectrum of the blood plasma sample from a patient with nasopharyngeal cancer, obtained by mixing the plasma with Ag colloid, (*B*) the regular Raman spectrum of the same plasma sample without the silver sol, and (*C*) the background Raman signal of the anticoagulant agent EDTA mixed with Ag colloid. Reprinted with permission from Ref. [34]. Copyright (2010) with permission from Elsevier

to detect the human radical S-adenosyl methionine domain containing 2 (RSAD2) RNA. The human RSAD2 gene is known to be a host-response biomarker for respiratory viral infections. The method is essentially a SERS effect modulation approach, associated with AgNP and Raman dye-labeled DNA hairpin probes. Hybridization with the target sequences opens the hairpin and spatially separates the Raman label from the silver surface, thus reducing the SERS signal of the label. The process is shown in Fig. 6.

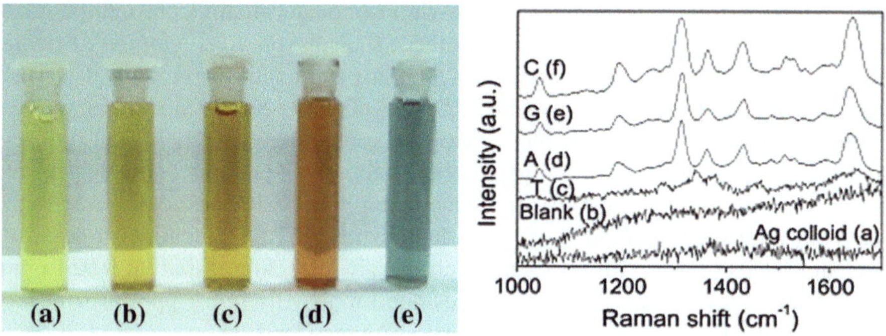

Fig. 5 Photographs of colloidal silver nanoparticles (**a**, *yellow–green*) and their aggregates induced by thymine (**b**, *yellow*), adenine (**c**, *orange*), guanine (**d**, *reddish–green*), and cytosine (**e**, *blue–green*) measured 3 min after the addition fig normalized (to 1,307 cm^{-1}) SERS spectra of 10^{-7} M RB for different DNA base-induced silver aggregates (*c–f*), normal Raman spectra of silver colloid (*a*), and blank RB at 10–7 M (*b*). Reprinted with permission Ref. [43]. Copyright (2008) with permission from Elsevier

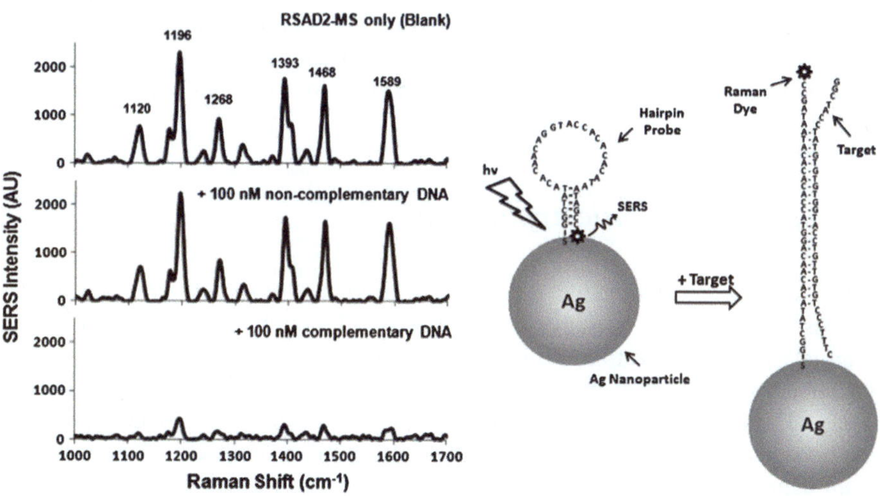

Fig. 6 Detection scheme of the molecular sentinel (MS) nanoprobes. Fig B.SERS spectra of the RSAD2-MS nanoprobes. *Upper spectrum* blank (no target DNA present). *Middle spectrum* in the presence of 100 nM non-complementary DNA (negative control). *Lower spectrum* in the presence of 100 nM complementary target DNA. Reprinted with permission from Ref. [47]. Copyright (2013) with permission from Elsevier

4.2 SERS for Detection of Toxic Substances

Another important application of SERS is to detect highly toxic molecules, such as chemical warfare agents and toxic industrial chemicals [48]. The key challenge here is to increase the binding affinity of the molecule with the nanoparticles, in order to observe SERS. Yan et al. reported SERS detection of nerve agent simulants such as methyl parathion, other chemical warfare simulants such as dimethyl methylphosphonate, pinacolyl methylphosphonate, diethyl phosphoramidate and 2-chloroethyl ethylsulfide, and bacteria such as *Bacillus globigii*, *Erwinia herbicola*, and *Bacillus thuringiensis*, which are simulants for biological warfare agents [49]. The high sensitivity of SERS, as well as its ease of operation and fast response, makes it highly suitable for environmental analysis. The recent progress of SERS in this area has highlighted the detection of organic pollutants, heavy metal ions and pathogens. A good, recent example of SERS pesticide detection is the work published by Yuan et al., who demonstrated SERS potential for sensitive and rapid detection of dithiocarbamate chemicals, detecting thiram at 0.024 ppm. This is much lower than the maximal residue limit (MRL) of 7 ppm in fruit prescribed by the U.S. Environmental Protection Agency [50].

The high sensitivity of SERS has also been demonstrated by several research groups in their work on the detection of pathogenic bacteria, such as Salmonella and Listeria. Recently, Zhou et al. synthesised Ag NPs on the cell walls of bacteria for label-free SERS detection of bacteria such as *E. coli* and *S. epidermidis* in drinking water [51]. Cowcher et al. reported using SERS for the detection of Bacillus [52]. In their study, silver colloidal substrates were used to detect Bacillus spores at 5 ppb, which is a highly desirable safety level in the food manufacturing industry. Melamine, a contaminant found in milk and toddler formula, was also detected by several groups using AgNP and SERS (for examples see [53–55]).

4.3 SERS for Forensic Applications

SERS has become technique ubiquitously used in forensic studies for various reasons, including its non-invasive nature, small sample size requirements and minimal sample preparation. SERS application for forensic analysis was first demonstrated by Seifar et al. in 2001 [56]. In their study, thirteen blue and thirteen black ink lines were distinguishable based on their Raman frequencies, spectral band shapes and relative band intensities, using silver colloid and poly-l-lysine as an aggregator with a 685 nm laser. The results correlated well with standard thin layer chromatography (TLC) used to distinguish inks. In another study, Leona et al., used a two-step methodology in which a methacrylate gel bead saturated with extracting solvents, such as dimethylformamide and ethylenediaminetetraacetic acid, was used to extract dye from coloured components. The extracted dye was then treated with SERS active AgNP for SERS analysis [57].

4.4 SERS for Single Molecule Detection

Single molecule identification and detection arguably represents the ultimate limit in biochemical analyses. As a single molecule tool, SERS offers exciting possibilities, as compared to traditional fluorescence, which is often encountered in single molecule study. The main drawbacks of using fluorescence for single molecule detection is that it does not provide detailed molecular information, and photobleaching often limits the number of photons obtainable from a single molecule. Due to the high structural information content of a Raman spectrum, SERS is an adequate technique for simultaneously detecting single molecules and identifying its chemical structure.

In 1997, two independent groups (Kneipp et al. and Nie et al.) reported single molecule detection using SERS. In her pioneering work, Kneipp achieved single molecule detection using AgNP for the first time [58]. The research demonstrated that by using extremely large, effective cross-sections available from surface-enhanced Raman scattering (SERS) and aggregating nanoparticles, a single crystal violet molecule in aqueous colloidal silver solution can be detected. The results showed the expected Poisson distribution for actually measuring 0, 1, 2 or 3 molecules when the scattering volume is 30 pl. Also in 1997, Nie and Emory demonstrated single R6G molecules adsorbed on citrate-reduced AgNP that were electrostatically immobilized on glass in an ambient environment by exploiting the surface enhanced Raman and resonance enhancement effects [59]. The argument presented for single molecule SERS is that the number of colloidal particles far exceeds the number of analyte molecules. Assuming a random Poisson distribution, the probability of finding more than one molecule on a single particle is extremely small. Thus, the surface-enhanced Raman signals observed from a single nanoparticle should correspond to a single analyte molecule [60].

One of the greatest potential applications of single molecule SERS is in the rapid sequencing of DNA at the single molecule level. Sun et al., synthesized a controllable AgNP aggregate system by using ethanol and cetyltrimethylammonium bromide (CTAB) capped AgNP [42]. The technique was used to detect two common probe molecules, R6G and 4-aminothiophenol (4-ATP) at the single molecule level. In another study, Braun et al. demonstrated a single molecule SERS sensing technique for the detection of RNA and DNA, in which AgNP were assembled on an Ag film via the complementary target DNA strand, creating a strongly enhanced Raman signal [61]. Over the last decade, there has been considerable interest in SERS single molecule detection of biomolecules such as hemoglobin, glucose, cancer genes, pathogens, biological toxins and viruses [62–66]. In one application of SERS single molecule detection, Xu et al. were able to distinguish single hemoglobin (Hb) protein molecules attached to isolated and immobilized AgNP. In their work, AgNP were prepared by a citrate reduction method at a concentration of 35 pM. The colloidal solution was incubated with a 10 pM solution of human adult met-Hb, which renders an average of 0.3 Hb molecules per Ag particle [62].

5 SERS Based Micro-Devices

SERS measurements are generally performed by exciting a sample in a cuvette with a monochromatic laser light. Cuvettes may be circular, square or rectangular, and are usually made of Quartz as it will transmit both incident and emitted Raman light. In cuvette based geometry, the free-space laser beam is tightly focused on the sample by an objective lens, resulting in a high intensity near the focal point within the sample-containing cuvette (Fig. 7). As the effective cross-sectional area is smaller (or has a higher focused intensity), a shorter effective interaction length L_{eff} is achieved and the two properties counterbalance one other. Accordingly, cuvette geometry, which is diffraction-limited for tighter focus of laser beam, is inefficient for increasing the effect of matter-light interaction [67, 68].

Researchers have used dielectric capillaries, or metal-coated tubes, to increase the effective interaction length [69]. Lee et al. demonstrated an Ag-coated capillary by synthesizing Ag sol through the reaction of $AgNO_3$ with poly-(ethylenimine) (PEI), toluene and benzenethiol (BT) [70]. The mixture was passed through a glass capillary to obtain a SERS-active Ag film-coated glass capillary. The group reported a detection limit of 1.0×10^{-7} M, and an S/N ratio of 3. In another study, Li et al. demonstrated using silver coated capillaries to reach a detection limit in the range of nM [71]. Using capillaries increases the effective interaction length, but a key disadvantage is their high propagation losses, as shown in Fig. 8 [72].

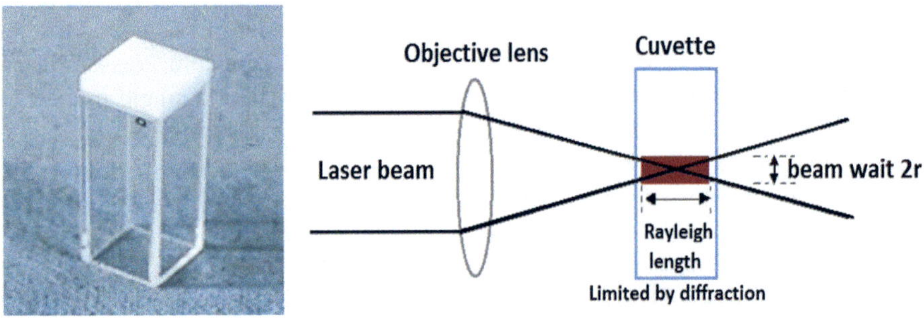

Fig. 7 Cuvette based geometry with L_{eff} limited by the Rayleigh length in a focused free-space laser

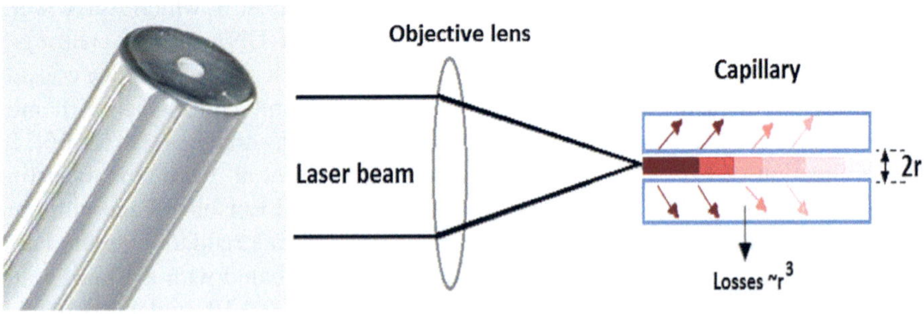

Fig. 8 Capillary based geometry with high propagation losses

An ideal configuration for effective interactions requires a diffraction-free, lossless, single-mode waveguide, with a core diameter equal to the focused laser beam waist. HC-PCF comes close to accomplishing this [73].

5.1 SERS Using HCPCF

The hollow core photonic crystal fiber (HC-PCF) is a novel type of optical waveguide that allows the target sample and nanoparticles to be infiltrated into the hollow core and cladding. HC-PCF provides an increase in interaction length with very low losses compared to cuvettes or capillaries, achieving high spatial confinement of the target molecules, with respect to nanomaterials. The light guidance in HC-PCF occurs via a photonic band gap, which delivers strong light-matter interaction and enhancement of the Raman signal.

HC-PCF has several major advantages over conventional sample cells [73–76], as follows:

- Low-waveguide losses of a few dB/m allow the use of long optical-path lengths and greatly enhance the effective light-matter interaction, as shown in Fig. 9.
- A sample under analysis typically uses at least a milliliter of volume when examined with test tube or cuvette. In contrast, HC-PCF uses samples in the nano- to picoliter range, significantly decreasing the sample consumption rate.
- The small required sample volume and large overlap of the propagating laser mode field with the sample in HC-PCF provide the potential to develop simple, compact and sensitive biosensors. At comparable input power, the intensity in the hollow core is five orders of magnitude higher than with cuvette-based approaches.
- PCFs are easily fabricated from chemically inert, high-quality silica glass, with negligible scattering, absorbance or fluorescence.
- Finally, PCF can demonstrate the potential of non-invasive and label free detection of biomolecules

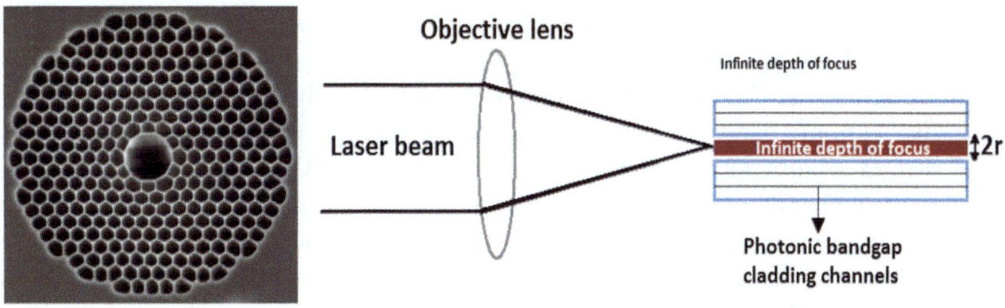

Fig. 9 HC-PCF guiding mechanism where high interaction length can be obtained due to the photonic bandgap that confines the light to propagate inside the core, thereby limiting loss

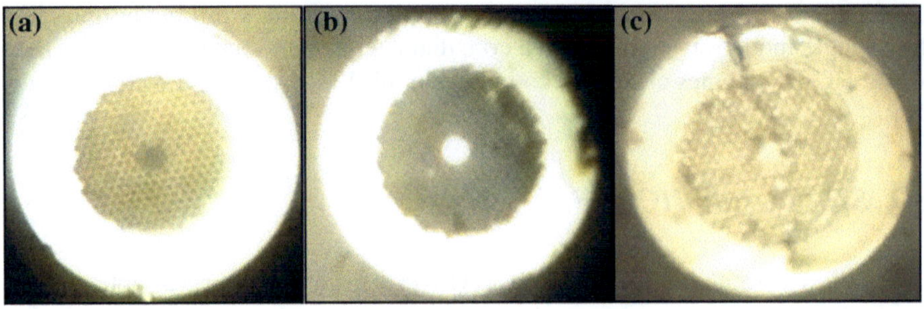

Fig. 10 *Cross-sectional view* of HC-PCF when **a** empty; **b** selectively filled; and **c** non-selectively filled

Over the last decade, there have been a number of publications that support the strong sensing platform offered by HC-PCF and SERS. As shown in Fig. 10, SERS in HC-PCF can be achieved in two ways [77–83]. The first is to fill only the core while sealing off the cladding holes (i.e. the selective filling approach), as demonstrated by Zhang et al. using a test molecule, R6G [84]. This was mixed with a AgNP colloidal solution, and a 10 mM concentration of NaCl was added to aggregate the nanoparticles for optimal SERS performance. With a 10 cm HC-PCF an enhancement of ~100 times was reported in the SERS experiment, with a lowest detectable concentration of 10^{-5} M.

In another demonstration, HC-PCFs were used as a SERS probe by coating AgNP on the inner wall of core and cladding, while non-selectively filling sample into HC-PCFs [82]. Han et al. showed SERS measurement in HC-PCF by functionalizing AgNP only in the core region, while sealing off the cladding holes [85]. In their work, light propagated in the sample filled core region through total internal reflection (index guided), and thereby generated the SERS signal of the sample. In another demonstration, Irizar et al., selectively filled the core of HC-PCF, and obtained enhancement in the Raman signal for studying different stages in the synthesis nanoparticles [86]. In the selective filling approach, the light is guided by total internal reflection (TIR) rather than the photonic bandgap effect, which does not use the full light guiding potential of photonic bandgap property of HC-PCF for. This method is quite cumbersome.

The second method of using HC-PCF for characterizing samples is to fill all the holes non-selectively. HC-PCF guiding at 800 nm (empty) was used by Yang et al. to guide a laser at 785 nm wavelength when filled with the sample [87]. In this case, the photonic band gap (or transmission wavelength) of the sample filled HC-PCF shifted significantly compared to the empty HC-PCF, and the shifted transmission wavelength of HC-PCF did not match the excitation wavelength. Thus, the light is only weakly guided into the HC-PCF and not tightly confined in the sample-filled core region, resulting in weak light-matter interaction; once again the photonic band gap is not preserved.

There is a method that does preserve the photonic bandgap and this also uses the non-selective filling approach [88–90]. By matching the excitation wavelength (785 nm) with that of the shifted transmission band maxima of the sample filled HC-PCF this is achieved. The interaction of the light and the sample is extremely strong, and results in a substantial enhancement of the Raman signal from the analyte. The light-guiding property of non-selectively filled HC-PCF changes in a manner that is dependent on the refractive index of the filled sample. In this case, although the guiding principle is still due to the bandgap effect, the transmission band supported by the fiber is shifted. The shift in the transmission wavelength of HC-PCF can be determined using the equation by Antonopoulos et al. [78]. This approach was also used by Gu's group to demonstrate various samples, such as R6G, human insulin and tryptophan, at the lowest detectable concentration of 10^{-10} M [88]. Dinish et al., demonstrated a practical approach for detecting cancer cells, using an ultrasensitive platform for the multiplex detection of cancer biomarkers by combining the SERS technique with a hollow-core photonic crystal fiber (HCPCF) [89]. The biomarkers were immobilized in the HCPCF using antibody-conjugated SERS-active nanoparticles (SERS nanotags). As a proof-of-concept for targeted multiplex detection, simultaneous detection of hepatocellular carcinoma (HCC) biomarkers, alpha fetoprotein (AFP) and alpha-1-antitrypsin (A1AT) using HC-PCF and SERS was successfully demonstrated.

5.2 *Silver Nanoparticles in HC-PCF for SERS*

This section describes recent work on the determination of the maximum sensing potential of a SERS based HC-PCF platform [90–92]. The study evaluates the contribution of HC-PCF and nanoparticles to the overall enhancement of the Raman signal of different biomolecules. It builds on previous work that established the role of HC-PCF as a Raman signal 'enhancer' due to its photonic band gap property [76, 77].

5.3 *HC-PCF Selection*

The light guiding property of non-selectively filled HC-PCF, changes depending on the refractive index of the filled sample. In this case, though the guiding principle is still due to the bandgap effect, the transmission band supported by the fiber is shifted. The shift in the transmission wavelength of HC-PCF can be determined from the following equation by Antonopoulos et al. [78]:

$$\lambda' = \lambda_0 \left[\frac{1 - \left(\frac{n_{liq}}{n_{sil}} \right)^2}{1 - \left(\frac{n_{air}}{n_{sil}} \right)^2} \right]^{1/2} \quad (5)$$

where λ_0 is the wavelength of the fiber when empty, λ' is the shifted wavelength of the fiber when filled with the sample, n_{liq} is the refractive index of the liquid sample, n_{air} is the refractive index of the air, and n_{sil} is the refractive index of the silica. Thus, depending on the excitation wavelength, fiber that could guide this wavelength when filled can be determined. With an excitation source of 785 nm, HC-1550 fiber is able to support the photonic bangap when filled with biological samples (n_{liq} ~ 1.33–1.37).

5.4 Enhancement of Raman Signal in HC-PCF with Rhodamine as a Test Molecule

This work focused on recording the Raman spectrum of a pure (R6G) solution filled into the HC-PCF [93]. For comparison, sample power in bulk (cuvette) solution was kept same as that in micro-litre (HC-PCF) solution. While comparing the Raman spectrum of the sample solution (R6G) in both these cases, one can draw the effective contribution of HC-PCF towards enhancing Raman signal of R6G. Figure 11 shows that the Raman signal of R6G obtained from HC-PCF is enhanced compared to that from the cuvette. The HC-PCFs are known to enhance the Raman signal, as it supports strong modal field overlap with the sample due to its photonic band gap property. The significance of this experiment was to determine the factor by which HC-PCF alone enhances the Raman signal, which we found to be ~90. Figure 11 shows the prominent Raman bands of R6G in the spectral range of 900–1,600 cm^{-1} and this study used the SERS peak at 1,365 cm^{-1} (aromatic C–C stretching) to evaluate the various enhancement factors that corresponded to nanoparticles and HC-PCF (see [94] for detailed assignment with respect to different vibrational modes).

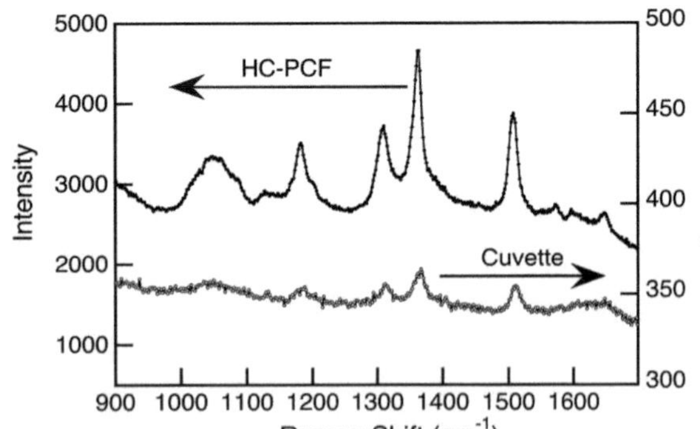

Fig. 11 Raman spectrum of pure R6G in HC-PCF and cuvette. R6G concentration was 10^{-3} M. All measurements were carried out in water at room temperature. Adapted from Ref. [93]

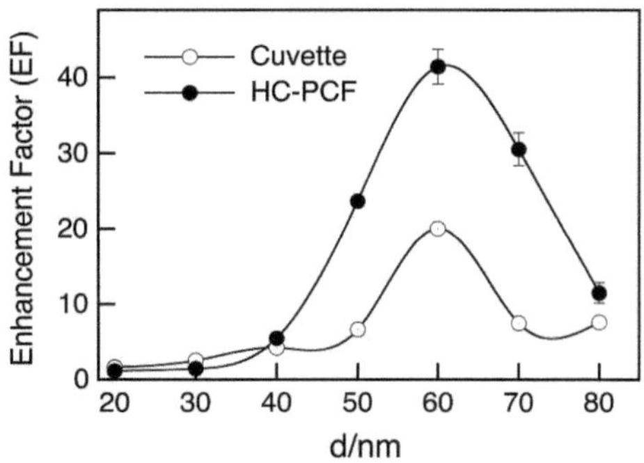

Fig. 12 Enhancement factor versus AgNP size as measured in HC-PCF and cuvette. *Spherical* shaped AgNP of different diameters were purchased from NanoComposix Inc. The silver concentration was 0.02 mg/ml in all cases. The sample volume ratio of AgNP and R6G was 1:2. Adapted from Ref. [93]

5.5 Optimal Size of AgNP for Maximum SERS Signal Enhancement

In order to determine the optimal size of AgNP, which will maximize the resulting SERS signal of R6G in HC-PCF, seven sample sets were prepared with identical volume ratios of NP and R6G, and varying AgNP diameters of 20, 30, 40, 50, 60, 70 and 80 nm.

Figure 12 shows that the EF HC-PCF values gradually increased with the increase of the AgNP size, and reached the maximum for AgNP at a diameter ~60 nm. When the AgNP size was increased beyond 60 nm, the EF HC-PCF decreased almost exponentially. Figure 12 also indicates that the error bar values for particles whose diameter was greater than 40 nm are relatively higher with an accuracy of 2–4 % compare to particles with diameter smaller than or equal to 40 nm. This is mainly due to higher scattering of light by bigger sized AgNP, thereby affecting the light guidance within HC-PCF. The optimal range of AgNP size for maximum SERS signal for bulk samples (cuvette) was 60 nm, which is consistent with our previous report [95].

5.6 SERS Enhancement in HC-PCF with Adenosine as a Test Molecule

Adenosine is considered a key molecule in clinical environments, owing to its role in regulating extra-cellular physiological activity. Monitoring adenosine is vitally important, due to its impact on pro inflammatory and tissue-destruction during bronchoconstriction in patients with asthma and chronic obstructive pulmonary disease (COPD) [96]. In this work, we explored an alternative method for monitoring adenosine that exploits surface enhanced Raman scattering (SERS) within hollow core photonic crystal fibers (HC-PCF).

Initially, the core and cladding holes of the HCPCF were filled with a sample solution of adenosine, and the SERS spectrum of adenosine and nanoparticles in the HC-PCF was recorded. It was then compared with the SERS spectrum of the

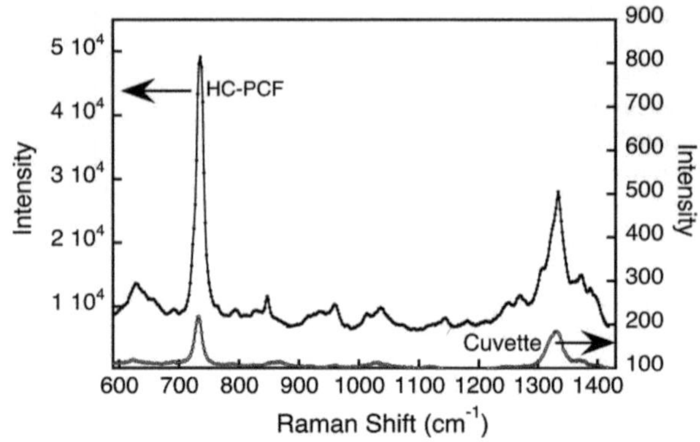

Fig. 13 Comparison of SERS spectrum of adenosine in HC-PCF and cuvette. Adenosine concentration of 25 mM was prepared in water at room temperature. Adapted from Ref. [93]

same sample mixture in a cuvette, and further compared with the normal Raman spectrum of adenosine in a cuvette, as shown in Fig. 13.

The results suggest that the HC-PCF mediates strong light-sample interaction via the photonic bandgap property, which results in enhancement of the Raman signal of adenosine by a factor of ~90, compared to that in a cuvette. Thus, HC-PCF clearly acted as a "Raman signal enhancer". When the SERS spectrum of the mix of adenosine and nanoparticles is compared with that of the normal Raman spectrum of pure adenosine in a cuvette, the contribution of nanoparticles to the enhanced Raman signal of adenosine was found to be ~20 times. Thus, our SERS based HC-PCF detection scheme yielded an overall enhancement of the Raman signal of adenosine by a factor of ~2,700 at 733 cm^{-1}, as shown in Fig. 13 [97].

To summarize, both nanoparticles and HC-PCF independently contribute to amplification of the Raman signal of adenosine. Hence, SERS and HC-PCF based sensing platforms have the potential to detect extremely low concentration of molecules of interest.

6 SERS Future and Challenges

6.1 Future

As discussed in this chapter, significant advances have been made in the use of SERS over the last 35 years. Single molecule SERS has been demonstrated for some biomolecules, and its potential to compete with fluorescence continues to be of great interest to researchers. There is also substantial effort to enhance SERS with Resonant Raman, by using tunable lasers as well as tuning the properties of the nanoparticles [43]. Investigation of new plasmonic materials, such as graphene semiconductors, is another promising area of future research [98–100].

The likelihood of using SERS to detect various chemicals has prompted exploration of its commercialization potential. A number of portable Raman spectrometers are now available, such as ReporteR, TruScan and Tellspec, and this has

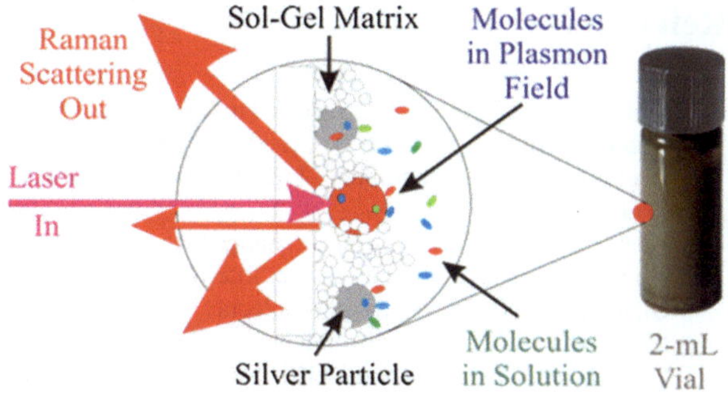

Fig. 14 Portable SERS Analyzer from Real Time Analyzer Inc. Reproduced with permission Real Time Analyzer Inc.® 2014

increased interest in developing commercial nanoparticle vials and substrates for SERS. Here, we present one example, the Portable SERS Analyzer (PSA), a Raman instrument that uses Real Time Analyzer's (RTA) patented SERS vials [101, 102]. This tool has been designed and optimized for trace chemical analysis, and in addition to SERS applications it can perform normal Raman measurements of unknown liquids and solids. Figure 14 shows a sample vial coated with a patented SERS-active sol-gel, as well as SERS Microplates. A solution containing the chemical of interest is simply injected into the vial, which is then placed in a Raman spectrometer sample compartment, and the instrument records the spectrum of the analyte. Other commercial products based on Ag nanoparticles for SERS applications are provided by nanoComposix, Sigma Aldrich, ThermoFisher, EnSpectr SERS Silmoco and VivNano.

6.2 Challenges

Though excellent results have been achieved with SERS its implementation does present some challenges, the greatest being the requirement of a high degree of reproducibility in the morphology and roughness of the metallic nanoparticles. This is problematic due to a number of factors, including temperature, time, concentration of the precursors, agitation speeding, dust and contamination [103], which can make SERS enhancements inconsistent and generally irreproducible. Another challenge is the tendency of metallic nanoparticles, including silver, to change their physiochemical properties (e.g. size and aggregation state, surface charge, surface chemistry, optical property) when they come in contact with complex biological fluids.

Acknowledgments This work was supported by Natural Sciences and Engineering Research Council of Canada (NSERC), Ontario Centre of Excellence (OCE), and The Technology Transfer and Business Enterprise (TTBE) at University of Ottawa.

References

1. Motz, J.T., et al.: In vivo Raman spectral pathology of human atherosclerosis and vulnerable plaque. J. Biomed. Opt. **11**(2), 021003–021003 (2006)
2. Campion, A., Kambhampati, P.: Surface-enhanced Raman scattering. Chem. Soc. Rev. **27**(4), 241–250 (1998)
3. Abalde-Cela, S., et al.: Surface-enhanced Raman scattering biomedical applications of plasmonic colloidal particles. J. R. Soc. Interface **7**(Suppl 4), S435–S450 (2010)
4. Qian, X.M., Nie, S.M.: Single-molecule and single-nanoparticle SERS: from fundamental mechanisms to biomedical applications. Chem. Soc. Rev. **37**(5), 912–920 (2008)
5. Oh, Y.J., et al.: Beyond the SERS: Raman enhancement of small molecules using nanofluidic channels with localized surface plasmon resonance. Small **7**(2), 184–188 (2011)
6. Choi, C.J., et al.: Surface-enhanced Raman nanodomes. Nanotech. **21**(41), 415301–415307 (2010)
7. Schatz, G.C., Van Duyne, R.P.: Electromagnetic mechanism of surface-enhanced spectroscopy. In: Handbook of Vibrational Spectroscopy. John Wiley & Sons, Inc., New York (2006)
8. Faulds, K., et al.: Comparison of surface-enhanced resonance Raman scattering from unaggregated and aggregated nanoparticles. Anal. Chem. **76**(3), 592–598 (2004)
9. McHugh, C.J., et al.: Selective functionalisation of TNT for sensitive detection by SERRS. Chem. Commun. **6**, 580–581 (2002)
10. McNay, G., et al.: Surface-enhanced Raman scattering (SERS) and surface-enhanced resonance Raman scattering (SERRS): a review of applications. Appl. Spectrosc. **65**(8), 825–837 (2011)
11. Graham, D., Goodacre, R.: Chemical and bioanalytical applications of surface enhanced Raman scattering spectroscopy. Chem. Soc. Rev. **37**(5), 883–884 (2008)
12. Esbensen, K.H. (ed.): An introduction to multivariate data analysis and experimental design. In: Multivariate Data Analysis–In Practice, 5th edn. Camo Inc., Oslo (2004)
13. Raman, C.V., Krishnan, K.S.: A new type of secondary radiation. Nature **121**, 501–502 (1928)
14. McCreery, R.L., Chemical analysis. In: Raman spectroscopy for chemical analysis, p. 448. John Wiley & Sons, Inc., New York (2000)
15. Singh, R.: C. V. Raman and the discovery of the Raman effect. Phys. Perspect. **4**(4), 399–420 (2002)
16. Fleischmann, M., Hendra, P.J., McQuillan, A.J.: Raman spectra of pyridine adsorbed at a silver electrode. Chem. Phys. Lett. **26**(2), 163–166 (1974)
17. Moskovits, M.: Surface-enhanced spectroscopy. Rev. Mod. Phys. **57**(3), 783–826 (1985)
18. Morton, S.M., Jensen, L.: Understanding the molecule-surface chemical coupling in SERS. J. Am. Chem. Soc. **131**(11), 4090–4098 (2009)
19. Jensen, L., Aikens, C.M., Schatz, G.C.: Electronic structure methods for studying surface-enhanced Raman scattering. Chem. Soc. Rev. **37**(5), 1061–1073 (2008)
20. Stiles, P.L., et al.: Surface-enhanced Raman spectroscopy. Ann. Rev. Anal. Chem. **1**(1), 601–626 (2008)
21. Stewart, M.E., et al.: Nanostructured plasmonic sensors. Chem. Rev. **108**(2), 494–521 (2008)
22. Kelly, K.L., et al.: The optical properties of metal nanoparticles: the influence of size, shape, and dielectric environment. J. Phys. Chem. B. **107**(3), 668–677 (2002)
23. Orendorff, C.J., et al.: Aspect ratio dependence on surface enhanced Raman scattering using silver and gold nanorod substrates. Phys. Chem. Chem. Phys. **8**(1), 165–170 (2006)
24. Orendorff, C.J., et al.: Surface-enhanced Raman spectroscopy of self-assembled monolayers: sandwich architecture and nanoparticle shape dependence. Anal. Chem. **77**(10), 3261–3266 (2005)
25. Schatz, G., Van Duyne, R.: Electromagnetic mechanism of surface enhanced spectroscopy. In: Chalmers, J.M. (ed.) Handbook of Vibrational Spectroscopy. Wiley, Chichester (2002)

26. Baker, G.A., Moore, D.S.: Progress in plasmonic engineering of surface-enhanced Raman-scattering substrates toward ultra-trace analysis. Anal. Bioanal. Chem. **382**(8), 1751–1770 (2005)
27. Zhang, X., et al.: Rapid detection of an anthrax biomarker by surface-enhanced raman spectroscopy. J. Am. Chem. Soc. **127**(12), 4484–4489 (2005)
28. Bogue, R.: Nanosensors: a review of recent progress. Sensor Rev. **28**(1), 12–17 (2008)
29. Liang, H., et al.: Highly surface-roughened "Flower-like" silver nanoparticles for extremely sensitive substrates of surface-enhanced Raman scattering. Adv. Mater. **21**(45), 4614–4618 (2009)
30. Baia, M., Baia, L., Astilean, S.: Gold nanostructured films deposited on polystyrene colloidal crystal templates for surface-enhanced Raman spectroscopy. Chem. Phys. Lett. **404**(1–3), 3–8 (2005)
31. Zhang, R., et al.: Highly efficient SERS test strips. Chem. Commun. **48**(47), 5913–5915 (2012)
32. El-Ansary, A., Faddah, L.M.: Nanoparticles as biochemical sensors. Nanotechnol. Sci. Appl. **3**, 65–76 (2010)
33. Sharma, B., et al.: SERS: materials, applications, and the future. Mater. Today **15**(1–2), 16–25 (2012)
34. Feng, S., et al.: Nasopharyngeal cancer detection based on blood plasma surface-enhanced Raman spectroscopy and multivariate analysis. Biosens. Bioelectron. **25**(11), 2414–2419 (2010)
35. Lin, J., et al.: A novel blood plasma analysis technique combining membrane electrophoresis with silver nanoparticle-based SERS spectroscopy for potential applications in noninvasive cancer detection. Nanomedicine **7**(5), 655–663 (2011)
36. Feng, S., et al.: Study on gastric cancer blood plasma based on surface-enhanced Raman spectroscopy combined with multivariate analysis. Sci. China Life Sci. **54**(9), 828–834 (2011)
37. Gong, J.-L., et al.: Ag/SiO2 core-shell nanoparticle-based surface-enhanced Raman probes for immunoassay of cancer marker using silica-coated magnetic nanoparticles as separation tools. Biosens. Bioelectron. **22**(7), 1501–1507 (2007)
38. Li, D., et al.: Label-free detection of blood plasma using silver nanoparticle based surface-enhanced Raman spectroscopy for esophageal cancer screening. J. Biomed. Nanotechnol. **10**(3), 478–484 (2014)
39. Girish, C.M., et al.: Rapid detection of oral cancer using Ag-TiO2 nanostructured surface-enhanced Raman spectroscopic substrates. J. Mat. Chem. B. **2**(8), 989–998 (2014)
40. Yang, J., et al.: Distinguishing breast cancer cells using surface-enhanced Raman scattering. Anal. Bioanal. Chem. **402**(3), 1093–1100 (2012)
41. Majumdar, D., et al.: DNA-mediated wirelike clusters of silver nanoparticles: an ultrasensitive SERS substrate. ACS Appl. Mater. Interfaces **5**(16), 7798–7807 (2013)
42. Sun, L., et al.: Ethanol-induced formation of silver nanoparticle aggregates for highly active SERS substrates and application in DNA detection. J. Phys. Chem. C **112**(5), 1415–1422 (2008)
43. Basu, S., et al.: Interaction of DNA bases with silver nanoparticles: assembly quantified through SPRS and SERS. J. Colloid Interface Sci. **321**(2), 288–293 (2008)
44. Harper, M.M., et al.: Detection of SERS active labelled DNA based on surface affinity to silver nanoparticles. Analyst **137**(9), 2063–2068 (2012)
45. Gao, F., Lei, J., Ju, H.: Label-free surface-enhanced Raman spectroscopy for sensitive DNA detection by DNA-mediated silver nanoparticle growth. Anal. Chem. **85**(24), 11788–11793 (2013)
46. Murphy, S., Huang, L., Kamat, P.V.: Reduced graphene oxide-silver nanoparticle composite as an active SERS material. J. Phys. Chem. C **117**(9), 4740–4747 (2013)
47. Wang, H.-N., et al.: Surface-enhanced Raman scattering molecular sentinel nanoprobes for viral infection diagnostics. Anal. Chim. Acta **786**(1), 153–158 (2013)
48. Pearman, W.F., Fountain, A.W.: Classification of chemical and biological warfare agent simulants by surface-enhanced Raman spectroscopy and multivariate statistical techniques. Appl. Spectrosc. **60**(4), 356–365 (2006)

49. Yan, F., Stokes, D.L., Wabuyele, M.B., Griffin, G.D., Vass, A.A., Vo-Dinh, T.: Surface-enhanced Raman scattering (SERS) detection for chemical and biological agents. In: Biomedical Optics 2004 (pp. 302–308). International Society for Optics and Photonics, (2004)
50. Yuan, C., et al.: Single clusters of self-assembled silver nanoparticles for surface-enhanced Raman scattering sensing of a dithiocarbamate fungicide. J. Mater. Chem. **21**(40), 16264–16270 (2011)
51. Zhou, H., et al.: SERS detection of bacteria in water by in situ coating with Ag nanoparticles. Anal. Chem. **86**(3), 1525–1533 (2014)
52. Cowcher, D.P., Xu, Y., Goodacre, R.: Portable, quantitative detection of bacillus bacterial spores using surface-enhanced Raman scattering. Anal. Chem. **85**(6), 3297–3302 (2013)
53. Ping, H., et al.: Visual detection of melamine in raw milk by label-free silver nanoparticles. Food Control **23**(1), 191–197 (2012)
54. Sun, F., et al.: Analytical methods and recent developments in the detection of melamine. TrAC Trends Anal. Chem. **29**(11), 1239–1249 (2010)
55. Kim, A., et al.: Melamine sensing in milk products by using surface enhanced Raman scattering. Anal. Chem. **84**(21), 9303–9309 (2012)
56. Seifar, R.M., et al.: Applicability of surface-enhanced resonance Raman scattering for the direct discrimination of ballpoint pen inks. Analyst **126**(8), 1418–1422 (2001)
57. Leona, M.: Microanalysis of organic pigments and glazes in polychrome works of art by surface-enhanced resonance Raman scattering. PNAS **106**(35), 14757–14762 (2009)
58. Kneipp, K., et al.: Single molecule detection using surface-enhanced Raman scattering (SERS). Phys. Rev. Lett. **78**(9), 1667–1670 (1997)
59. Nie, S., Emory, S.: Probing single molecules and single nanoparticles by surface-enhanced Raman scattering. Science **275**(5303), 1102–1106 (1997)
60. Kneipp, K., Kneipp, H.: Single molecule Raman scattering. Appl. Spectrosc. **60**(12), 322A–334A (2006)
61. Braun, G., et al.: Surface-enhanced raman spectroscopy for DNA detection by nanoparticle assembly onto smooth metal films. J. Am. Chem. Soc. **129**(20), 6378–6379 (2007)
62. Xu, H., et al.: Spectroscopy of single hemoglobin molecules by surface enhanced Raman scattering. Phys. Rev. Lett. **83**(21), 4357–4360 (1999)
63. Chen, H., et al.: Investigation of the synthesis, SERS performance and application in glucose sensing of hierarchical 3D silver nanostructures. New J. Chem. **38**(8), 3907–3916 (2014)
64. Huh, Y., Chung, A., Erickson, D.: Surface enhanced Raman spectroscopy and its application to molecular and cellular analysis. Microfluid. Nanofluid. **6**(3), 285–297 (2009)
65. Premasiri, W.R., et al.: Characterization of the surface enhanced raman scattering (SERS) of bacteria. J. Phys. Chem. B. **109**(1), 312–320 (2005)
66. Shanmukh, S., et al.: Rapid and sensitive detection of respiratory virus molecular signatures using a silver nanorod array SERS substrate. Nano Lett. **6**(11), 2630–2636 (2006)
67. Siesler, H.W., Ozaki, Y., Kawata, S., Heise, H.M.: Near-infrared spectroscopy: principles, instruments, applications, p. 361. Wiley-VCH Verlag GmbH, Weinheim
68. Smith, E., Dent, G.: Modern Raman Spectroscopy—A Practical Approach. John Wiley & Sons, Inc., San Francisco (2005)
69. Marcatili, E., Schmeltzer, R.: Hollow metallic and dielectric waveguides for long distance optical transmission and lasers (long distance optical transmission in hollow dielectric and metal circular waveguides, examining normal mode propagation). Bell Syst. Tech. J. **43**, 1783–1809 (1964)
70. Lee, J.W., et al.: Novel fabrication of silver-coated glass capillaries for ready SERS-based detection of dissolved chemical species. Anal. Bioanal. Chem. **397**(2), 557–562 (2010)
71. Li, Y.S., Lee, A.S.: Silver-coated capillary tube for surface-enhanced Raman scattering. Vib. Spectros. **3**(2), 115–120 (1992)
72. Trutna, W.R., Byer, R.L.: Multiple-pass Raman gain cell. Appl. Opt. **19**(2), 301–312 (1980)

73. Thang, N.: Thesis: Stimulated Raman Scattering in Gas Filled Hollow-Core Photonic Crystal Fiber. Max Plack Institute, Germany (2013)
74. Cox, F.M., Argyros, A., Large, M.C.J.: Liquid-filled hollow core microstructured polymer optical fiber. Opt. Express **14**(9), 4135–4140 (2006)
75. Khetani, A.: Thesis: Photonic Crystal Fiber as a Biosensor. University of Ottawa, Ottawa (2008)
76. Khetani, A., et al.: Monitoring of heparin concentration in serum by Raman spectroscopy within hollow core photonic crystal fiber. Opt. Express **19**(16), 15244–15254 (2011)
77. Khetani, A., et al.: A method for using photonic crystal fiber as a Raman biosensor. US Patent-2010/0014077, USA (2010)
78. Antonopoulos, G., et al.: Experimental demonstration of the frequency shift of bandgaps in photonic crystal fibers due to refractive index scaling. Opt. Express **14**(7), 3000–3006 (2006)
79. Yan, H., et al.: Hollow core photonic crystal fiber surface-enhanced Raman probe. Appl. Phys. Lett. **89**(20), 204101 (2006)
80. Jensen, J.B., et al.: Photonic crystal fiber based evanescent-wave sensor for detection of biomolecules in aqueous solutions. Opt. Lett. **29**(17), 1974–1976 (2004)
81. Nielsen, K., et al.: Selective filling of photonic crystal fibres. J. Opt. A. **7**(8), L13–L20 (2005)
82. Shi, C., et al.: Inner wall coated hollow core waveguide sensor based on double substrate surface enhanced Raman scattering. Appl. Phys. Lett. **93**(15), 153101–153101-3 (2008)
83. Huang, Y., Xu, Y., Yariv, A.: Fabrication of functional microstructured optical fibers through a selective-filling technique. Appl. Phys. Lett. **85**(22), 5182–5184 (2004)
84. Zhang, Y., et al.: Liquid core photonic crystal fiber sensor based on surface enhanced Raman scattering. Appl. Phys. Lett. **90**(19), 193504 (2007)
85. Han, Y., et al.: Towards full-length accumulative surface-enhanced raman scattering-active photonic crystal fibers. Adv. Mater. **22**(24), 2647–2651 (2010)
86. Irizar, J., et al.: Raman spectroscopy of nanoparticles using hollow-core photonic crystal fibers. IEEE J. Sel. Topics Quant. Electron. **14**(4), 1214–1222 (2008)
87. Yang, X., et al.: Photonics crystal fiber Raman sensors. In: SPIE-Information Optics and Optical Data Storage II. SPIE Proceedings, Beijing, China (2012)
88. Yang, X., et al.: Hollow-core photonic crystal fibers for surface-enhanced Raman scattering probes. Int. J. Opt. **2011**, 11 (2011)
89. Dinish, U.S., et al.: Sensitive multiplex detection of serological liver cancer biomarkers using SERS-active photonic crystal fiber probe. J. Biophotonics. doi:10.1002/jbio.201300084 (2014)
90. Khetani, A., et al.: Hollow core photonic crystal fiber as a reusable Raman biosensor. Opt. Express **21**(10), 12340–12350 (2013)
91. Tiwari, V.S., et al.: Detection of amino acid neurotransmitters by surface enhanced Raman scattering and hollow core photonic crystal fiber. In: Proceedings of SPIE 8233, Reporters, Markers, Dyes, Nanoparticles, and Molecular Probes for Biomedical Applications IV. San Francisco, California, USA (2012)
92. Khetani, A., et al. Monitoring of adenosine within hollow core photonic crystal fiber by surface enhanced Raman scattering (SERS). In: 2011 11th IEEE Conference on Nanotechnology (IEEE-NANO). Portland, OR (2011)
93. Tiwari, V.S., et al.: Optimum size and volume of nanoparticles within hollow core photonic crystal fiber. IEEE J. Sel. Topics Quantum Electron. **20**(3), 205–212 (2014)
94. Hildebrandt, P., Stockburger, M.: Surface-enhanced resonance Raman spectroscopy of Rhodamine 6G adsorbed on colloidal silver. J. Phys. Chem. **88**(24), 5935–5944 (1984)
95. Stamplecoskie, K.G., et al.: Optimal size of silver nanoparticles for surface-enhanced Raman spectroscopy. J. Phys. Chem. C **115**(5), 1403–1409 (2011)
96. Johnston, D.L., et al.: Pulmonary function monitoring during adenosine myocardial perfusion scintigraphy in patients with chronic obstructive pulmonary disease. Mayo Clin. Proc. **74**(4), 339–346 (1999)

97. Sanches-Cortes, S., Garcia-Ramos, J.V.: Surface-enhanced Raman spectroscopy of adenosine and 5'AMP: evolution in time. **22**, 819–824 (*1991*)
98. Harper, M.M., McKeating, K.S., Faulds, K.: Recent developments and future directions in SERS for bioanalysis. Phys. Chem. Chem. Phys. **15**(15), 5312–5328 (2013)
99. Xu, W., Mao, N., Zhang, J.: Graphene: a platform for surface-enhanced Raman spectroscopy. Small **9**(8), 1206–1224 (2013)
100. Wang, X., et al.: Surface-enhanced Raman scattering (SERS) on transition metal and semiconductor nanostructures. Phys. Chem. Chem. Phys. **14**(17), 5891–5901 (2012)
101. Real Time Analyzer Inc.: http://www.rta.biz/. Accessed 31 March 2009
102. Inscore, F.E., Gift, A.D., Maksymiuk, P., Farquharson, S.: Characterization of chemical warfare G-agent hydrolysis products by surface-enhanced Raman spectroscopy. In: Optics East (pp. 46–52). International Society for Optics and Photonics, (2004)
103. Kahraman, M., et al.: Reproducible surface-enhanced Raman scattering spectra of bacteria on aggregated silver nanoparticles. Appl. Spectrosc. **61**(5), 479–485 (2007)

Silver Nanoparticles in Heterogeneous Plasmon Mediated Catalysis

María González-Béjar

Abstract Metallic nanoparticles are considered a new class of heterogeneous photocatalysts due to their ability to absorb solar light and convert it into chemical energy. Here, we focus on the use of silver nanoparticles in light driven catalytic processes. This chapter will start with a brief introduction to the currently available nanocomposites that contain silver nanoparticles. Then, the mechanisms behind the photocatalytic activity of those nanocomposites will be discussed.

Keywords Photocatalysis · Heterogeneous plasmon mediated catalysis · Silver nanoparticles

1 Introduction

Light of the appropriate wavelength can selectively activate a desired photocatalyst to generate reactive species that can in turn initiate processes that cannot occur under thermal conditions or that require harsh reagents [1, 2]. Undoubtedly, solar light is considered the greenest reagent for catalytic processes (free and abundant). Photocatalysis is the acceleration of a chemical reaction or transformation produced thanks to the absorption of light by a photocatalyst.

Nowadays, photochemical and photophysical applications have become fundamental for our daily life products: information technology, nanotechnology, sustainable technologies (solar energy storage, waste water cleaning, photovoltaics), lighting (LEDs), etc. [3]. Specifically, photocatalytic applications include advanced oxidation processes (e.g. water decontamination of organic debris) [4, 5], water splitting [6–11], self-cleaning surfaces [12–14], etc.

M. González-Béjar (✉)
Instituto de Ciencia Molecular (ICMol)/Departamento de Química Orgánica,
Universidad de Valencia, C/Catedrático José Beltrán 2, 46980 Valencia, Paterna, Spain
e-mail: maria.gonzalez@uv.es

E.I. Alarcon et al. (eds.), *Silver Nanoparticle Applications*, Engineering Materials,
DOI 10.1007/978-3-319-11262-6_4

Metallic nanoparticles (MNPs) possess unique optical, electrical, mechanical, and chemical properties. Thus, their use as photocatalysts to aid in the development of green nanotechnology has attracted a great deal of interest [11, 15–17]. Size reduction increases the Fermi level of the nanoparticle leading to a lower reduction potential of the metal on its surface (see Chap. "Silver Nanoparticles: From Bulk Material to Colloidal Nanoparticles"). The total surface area in nanomaterials is considerably higher than the obtained in the bulk material at the same total concentration. This translates, for example, into higher catalytic performance [18], which is extremely important for catalytic and photocatalytic applications, since one nanoparticle can simultaneously interact with many reactants (comparatively, organic molecules or organometallic complexes only allow one to one interactions) [19].

MNPs upon interaction with an electromagnetic wave give rise to an absorption band known as Surface Plasmon Band (SPB). See Chap. "Silver Nanoparticles: From Bulk Material to Colloidal Nanoparticles" for a detailed explanation. Mahmoud et al. [20]. found that the most active MNPs for catalysis are those with sharp edges, sharp corners, or rough surfaces. The shape of the MNP influences the nature of the surface (facet) exposed to reactants. This would change how reactants and MNP interact and therefore, would influence the outcome of the catalytic processes [21].

This chapter focuses on the use of silver nanoparticles (AgNP) as catalysts. These NPs are also used for other applications such as surface enhanced Raman spectroscopy (Chap. "Surface Enhanced Raman Scattering (SERS) Using Nanoparticles"), single-molecule spectroscopy [22–24], solar cells [25, 26] and biological uses [27] (Chap. "Biomedical Uses of Silver Nanoparticles: From Roman Wine Cups to Biomedical Devices"). For AgNP, a shift in the plasmon absorption band from 400 to 670 nm is observed as the particle shape transitions from a sphere to a cube [11, 28] AgNP' shapes have been found to affect the outcome of the catalytic oxidation of styrene to styrene oxide [29, 30].

MNPs have recently started to be applied in photocatalysis (also called Plasmon Mediated Catalysis, PMC) [11, 31] The use of PMC in synthetic applications is fairly recent and considerable effort is being made to optimize photoreactions promoted by localized plasmons [11, 15–17]. In other words, MNPs are used to transform the energy of light into chemical energy by the formation of short-lived energetic electrons and enhanced electric fields that are generated on the surface and/or in close proximity to the NPs surface [11]. Remarkably, very high-intensity fields have been found in the regions (termed 'hot spots') between plasmonic nanostructures [11].

MNPs interact with the reactant molecules on their surface. Simultaneously, the electrons at their surface can be excited by light. The reaction rate therefore, will depend not only on the number and energy of those electrons, but also on the number of reactant molecules on the NP surface dictated by the affinity between the MNP surface and the reactants [15–17]. Interestingly, plasmonic nanostructures effectively couple thermal and photonic stimuli to drive chemical transformations [32].

For a detailed summary of the essential features of the preparation methods of supported AgNP, the reader is directed to a recent review written by Wang et al. and the references cited therein [17].

Similar to gold nanoparticles, AgNP have optical properties and catalytic capabilities that make them ideal for green photocatalysis (i.e. photocatalysis utilizing visible and ultraviolet light, a major part of the solar spectrum) [15–17]. An ideal green photocatalyst must absorb sunlight (since it is naturally available), must be stable under the reaction conditions (i.e. solvent, or, if there is no other choice, temperature and illumination), must be available, reusable, able to perform selective transformations, and have low toxicity [2].

Photocatalysis is currently used in two particularly beneficial albeit contrasting purposes; (i) pollutant reduction and the destruction of harmful organic pollutants in contaminated water and air using techniques collectively known as advanced oxidation processes (AOPs) [5], and (ii) the synthesis of chemical compounds, "positive photocatalysis", whose aim is to generate new products [33–37].

Several organic compounds, organometallic complexes and semiconductors have been used as photocatalysts in both homogeneous and heterogeneous systems [1, 2]. Colloidal heterogeneous nanocatalysis in particular, refers to those reactions that occur on the surface of nanoparticles [20]. The active site in heterogeneous catalytic reactions is a collection of surface atoms that adsorb reactants and facilitate chemical bond transformation [21]. The loading of nanoparticles on solid supports is a brilliant strategy that is broadening the way of thinking about heterogeneous thermal catalysis and PMC. This chapter will focus on this strategy and its use in PMC.

There are several advantages in using heterogeneous PMC;

(i) MNPs have much better affinity than solid supports to many reactants, especially organic molecules [15].
(ii) MNPs are more resistant to degradation than organic photocatalysts.
(iii) Heterogeneous photocatalysts are easier to remove from the reaction mixture (filtration, centrifugation) than heterogeneous catalysts. This heterogeneity can favor the reaction of one kind of reactant in the presence of others by selective adsorption [1].
(iv) The MNPs get energy from the irradiation source, resulting in high-energy active surface atoms, which is desirable for activating molecules for chemical reactions.
(v) The predictable aggregation of the NPs is prevented by the interaction with the solid support [15].

There are several issues that need to be resolved for the successful application of PMC;

(i) Reshaping of the nanocatalyst during catalysis reaction [21].
(ii) Reaction products deposited on the NP surface could decrease NP stability and limit their recycling.
(iii) The reaction rate may be adversely affected due to the different states of matter in the mixture, or inefficient reactant adsorption/desorption processes, on the solid photocatalysts.

(iv) Reduced availability of active surface area due to NP interaction with the solid support.
(v.) The synthesis of MNPs usually involves reducing agents, organic cappings, and/or growth-directing molecules. Some of those compounds could end up attached to the particle, and thus affect the catalytic process itself [21].
(vi) Particularly in the case of AgNP, it is difficult to prevent the oxidation of AgNP to silver oxide [38].

Despite the above unresolved issues, recent work has demonstrated a plasmonically active and self-regenerative Ag/AgO_x nanostructure to be useful for catalytic reactions and this may open up a new perspective in the field [39]. The support-free catalyst shows plentiful surface-adsorbed oxygen species along with excellent localized surface plasmon resonance (LSPR) and measurable photoluminescence [39].

2 Heterogeneous Photocatalysts Based on AgNP

Semiconductors and insulator supports with surface-embedded metallic nanoparticles for improved stability are being developed and tested. Interestingly, the support can also help in some cases to perform the photocatalytic process, thanks to a synergistic effect generated by the association between the support and the nanoparticle. Accordingly, a thorough understanding of the properties of the new material is required in order to prepare the most efficient photocatalyst possible. Based on the above, research has been carried out on the photocatalytic activity of silver nanoparticles supported on titanium ($AgNP@TiO_2$) [38, 40–44], and other metal oxides (e.g. ZrO_2, SiO_2 and graphene oxide), as well as zeolites [45]. Other metal–semiconductor composite photocatalysts based on the Ag@AgX (X = Cl, Br, I) system have also been developed [17].

It is reasonable to expect different mechanisms for insulator supported AgNP and semiconductor-supported AgNP when applied to photocatalytic processes [15–17]. In this section, we describe these mechanisms in detail to afford the reader with a more in-depth understanding of the photophysical processes required to determine the appropriate strategy for their particular needs. While reviewing these mechanisms, the use of AgNP-based photocatalysis for advances oxidation processes will be briefly summarized and following that, a section focused on positive photocatalysis is presented.

2.1 Heterogeneous Photocatalysts Composed of AgNP and Semiconductors

Semiconductor photocatalysts based on TiO_2 are widely studied for applications in environmental remediation and solar energy conversion [46–49]. Unfortunately,

TiO_2 can only be excited by ultraviolet (UV) light due to its relative wide band gap (3.2 eV). Various strategies have however been developed to make it sensitive to visible light [46–48, 50–57].

Upon illumination of a semiconductor with light, one electron from the conduction band (CB) is promoted to the valence band (VB), leaving behind a vacancy, commonly known as a hole, in the CB [6]. The energy carried by photons from the illumination source has to be greater than the bandgap energy, E_g, (the ΔE between the CB and the VB) for this process to occur. These electron-hole pairs can recombine or be trapped and, most importantly, the photogenerated charge carriers can migrate to the photocatalyst surface and a redox process can take place, giving rise to the corresponding reduced and oxidized products. Some semiconductors respond to visible light (e.g. CdS), allowing scientists to use sunlight or indoor lighting to excite them [2, 58]. Admittedly, the use of photocatalysts based on heavy metals is not the ideal scenario for water and air purification. Thus, in order to overcome this impropriety, several metal-free quantum dots are being developed as potential photocatalysts [59].

A particularly effective strategy for narrowing the band gap of TiO_2 is called defect engineering [55, 60–64]. Introducing oxygen vacancies into the lattice of TiO_2, the light absorption of TiO_2 can be extended to the visible or even infrared region [61, 65]. It is equally important to optimize the photogenerated electron/hole separation characteristics over the TiO_2 surface. Notably, the addition of noble metals to TiO_2 could reduce the photogenerated electron/hole recombination [50, 51, 66].

AgNP supported on TiO_2 (AgNP@TiO_2) have shown photocatalytic activity under UV and visible light and also antibacterial properties [38, 40–43, 67–73]. The ability of AgNP@TiO_2 nanoparticles to degrade phenol [43, 74], methyl red [40], methyl orange (MO) [67] and methylene blue (MB) [38] has been evaluated. An electron transfer reaction takes place upon UV irradiation (Scheme 1) and, where spherical AgNP coated with TiO_2 are concerned, a spectral shift of the plasmon resonance band to a shorter wavelength was detected and elucidated to be due to electron trapping on the AgNP induced by the charge separation after the electron transfer process [40–42]. The interfacial charge transfer between TiO_2 and silver could be enhanced by a negative shift in the Fermi level of the AgNP@TiO_2 composite resulting from the accumulation of electrons produced by the surface plasmon resonance [16].

If an electron acceptor such as thionine dye or oxygen is present, the photogenerated electrons are scavenged and electrons fail to accumulate in the AgNP core [42].

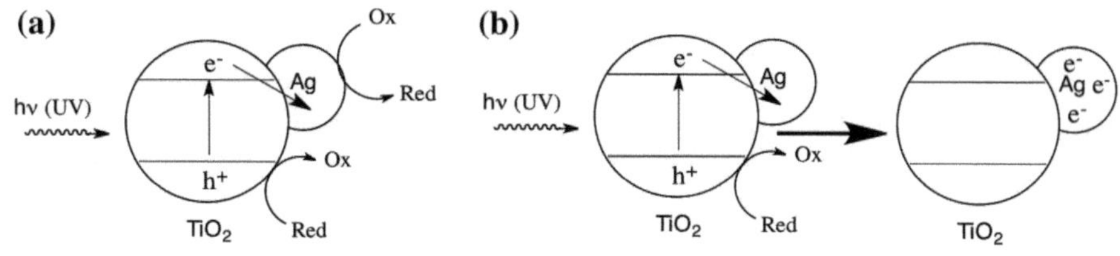

Scheme 1 Proposed mechanism for UV excitation of the AgNP@TiO_2 photocatalyts: **a** semiconductor excitation and electron transfer to AgNP [42, 43] and **b** photoinduced charge separation and charging of the AgNP [41]

This is also the case for Ag@ZnO [75–79]. It is in fact possible to carry out a redox titration and obtain quantitative information about the stored electrons in the Ag@TiO_2 colloids [42]. The capacity of electron storage is determined by the size of the metal core and its ability to undergo charge equilibration with the TiO_2 shell. Once this maximum storage limit is attained, electron-hole recombination in the TiO_2 shell dominates [42].

Electrons accumulated in AgNP can lead to generation of $^{\bullet}O_2^-$ when reacting with O_2, and finally to $^{\bullet}OH$. On the other hand, photogenerated holes in TiO_2 can be scavenged by OH^- or H_2O to generate $^{\bullet}OH$ directly [4, 80]. Moreover, the pH value also dictates the amount of $^{\bullet}OH$ generated [40] and the acidity of the hydroxyl groups on the TiO_2 surface could be enhanced in the presence of AgNP (as described for gold nanoparticles [81]). Thus, AgNP reduce electron-hole pairs recombination and enhance the efficiency of photoreduction owing to the Fermi level equilibration, ultimately leading to a higher yield of $^{\bullet}OH$ [40]. Indeed, due to the electronegativity of AgNP, their Fermi level can shift to negative potentials, resulting in charging effects or original chemical reactivity [82].

Unfortunately, the photo-oxidation of AgNP by O_2 takes place under visible light irradiation of a Ag@TiO_2 film [83] and the photogenerated holes residing on the electron-rich AgNP, drive dissolution of Ag atoms from the AgNP via Ag^+ ejection [84–86]. This can be prevented by either the deposition of TiO_2 made on AgNP, covered with a silica (SiO_2) shell [38], or the preparation of Ag@TiO_2 photocatalyst using Montmorillonite as a support [87]. Either approach effectively prevents the loss of the metal particles during the photocatalytic reaction.

Recently, metal-core@semiconductor-shell nanoparticles consisting of Ag@Cu_2O core-shell were shown to have photocatalytic activity where degradation of MO was concerned. This activity is due to the presence of localized surface plasmon resonance (LSPR) in the AgNP core [88]. This strategy avoids corrosion and dissolution of the metal particles and maximizes the metal—support interaction, thereby facilitating the plasmonic energy transfer processes. Lastly, the local electromagnetic field of the LSPR penetrates the shell, which can be used to tune the center wavelength of the LSPR by changing the shell thickness [88].

2.2 Heterogeneous Photocatalysts Composed of AgNP and Insulator Supports

In 2008, Awazu et al. demonstrated that the photolysis of methylene blue (organic dye) was promoted by a plasmonic effect when visible light was irradiated onto AgNP with a SiO_2 layer placed inside a TiO_2 layer whose plasmon resonance band was near the band edge of TiO_2 [38].

Very recently, AgNP were loaded onto inert supports, which are photocatalytically inactive supports. In this scenario, the charge density is partially localized on the AgNP surface, and this localization is increased by charge separation derived from the LSPR effect [89].

Insulators have different properties than semiconductors since the forbidden band gap between the valence band and conduction band is larger in an insulator, and as a result to this, electrons cannot be promoted from the valence band to the conduction band, even by near UV light [15].

Consequently, when AgNP are deposited onto an insulating oxide, there is no electron transfer between the photo-excited AgNP and the insulator support, and hence, only the nanoparticles participate in the photocatalysis directly [17]. In this case, the photocatalytic activity of noble metal NPs is determined by two factors: the enhanced electromagnetic field of neighboring NPs and the direct interaction of the LSPR excited energetic electrons with reactants. A distinct feature of these systems is that both light absorption and activation of the reactants take place on the AgNP. Charge transfer between the AgNP and support is not required. The particles can be dispersed on an insulating solid (or a very wide band gap semiconductor, such as Al_2O_3 (band gap, 8.0 eV) [90], ZrO_2 [45], SiO_2 (band gap, 9.0 eV) [45, 91], SBA-15 [92], zeolite [45], etc.), which has large specific surface areas and porosity.

The energy of incoming photons is concentrated into small volumes surrounding the AgNP. In fact, a build-up of intense, spatially non-homogeneous oscillating electric fields takes place in the neighborhood of the nanoparticle in the AgNP@insulator nanostructure due to the localize plasmon resonance of AgNP [23, 25, 93]. Hot spots also display very high-intensity fields [94] and these enhanced electric fields play a key role in photocatalysis. Indeed, for nanoparticles smaller than 30 nm, surface plasmons lead to the formation of energetic charge carriers [93, 95], which can be transferred to the surroundings [96] or relaxed by locally heating the nanoparticle [26]. This mechanism is thus quite different from that which occurs in the presence of a semiconductor, where electron transfer between the AgNP and the semiconductor takes place.

AgNP supported on zirconia, silica and Zeolite Y (Ag@ZrO_2, Ag@SiO_2 and Ag@Zeolite Y) are active photocatalysts for sulforhodamine-B (SRB) and phenol degradation [45]. The degradation of SRB by these photocatalysts increases proportionally with silver content.

Also, Ag@SiO_2 exhibited superior photocatalytic performance toward MB photodegradation [91]. Once AgNP are excited, the generation of photo-excited electrons and electron vacancies takes place and are responsible for MB degradation (Scheme 2).

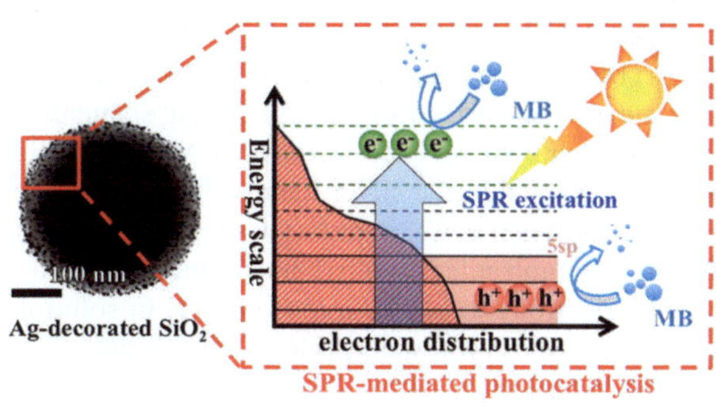

Scheme 2 Schematic diagram illustrating a AgNP@SiO_2 nanohybrid that display photocatalytic activity upon surface plasmon resonance (SPR) excitation and degrades methylene blue (MB). Reprinted with permission from Ref. [91]. Copyright 2012 American Chemical Society

Another interesting aspect of photocatalysts containing AgNP is the possible enhancement of the photocatalytic effect of an anchored molecule. For example, core–shell Ag@SiO_2 nanoparticles with an anchored $[Ru(bpy)_3]^{2+}$ dye, showed an enhancement of the photo induced oxidation activity of the ruthenium(II) complex that was attributed to the surface plasmon resonance [97].

2.3 Heterogeneous Photocatalysts Composed of AgNP Ag@AgX (X = Cl, Br)

Silver halides such as AgCl and AgBr are another class of semiconductor supports, which have been extensively used in photographic films due to their photosensitivity. The band gaps, E_g, of AgCl and AgBr are 3.25 and 2.69 eV, respectively [26, 38, 98–100]. Silver halides have been mainly used to host AgNP. Although, the synthesis of silver halide nanostructures has not been as well studied as that of AgNP [101], the synthesis of heterogeneous photocatalysts composed of AgNP Ag@AgX (X = Cl, Br, I) has been extensively investigated over the last decade.

The photocatalytic activity of Ag@AgCl nanoparticles fabricated with different morphological structures was recently shown to extend to visible light due to the SPB of AgNP [102–113]. Various morphotypes of the same NPs have been explored including cubes [106, 109, 112, 113], cubic cages [114], wires [103] films [110] and other shapes [115, 116]. Cubic Ag@AgCl have higher specific surface area, more active sites and facets as compared with spherical counterparts of similar size, all of which are favorable to the enhancement of photocatalytic activity [108]. However, enhanced photocatalytic activities have been reported when using red Ag@AgCl photocatalysts due to their even larger surface area [111].

AgNP supported on AgCl (Ag@AgCl) with high stability and high photocatalytic efficiency under illumination have been used to disinfect water and decompose organic pollutants such as phenol, and the dyes; 2,4-dichlorophenol, methyl orange (MO) [103, 107, 109–114], orange 7 [109], rhodamine B and methylene Blue [106]. These nanocomposites also show reductive activities such as, conversion of CO_2 into liquid carbon fuels (methanol and ethanol) [111, 115] and reduction of Cr^{VI} to Cr^{III} [111, 117].

Upon UV light excitation, a silver halide particle generates an electron-hole pair. The electrons migrate to the surface of the nanoparticle, while the holes migrate to the surface of the AgCl. Subsequently, the photogenerated electron combines with an Ag^+ ion to form an Ag^0 atom [107]. Meanwhile, the photogenerated holes within the silver halide photocatalysts are able to oxidize halide ions to halide atoms (i.e. Cl^- ions to Cl^0 atoms (radicals)), which are strongly oxidative for organic molecules, finally being reduced back to Cl^- ions [107, 109].

The electrons would be trapped by O_2 in solution to form superoxide ions and other reactive oxygen species (ROS) that could also promote the decomposition of dyes [4, 109, 118].

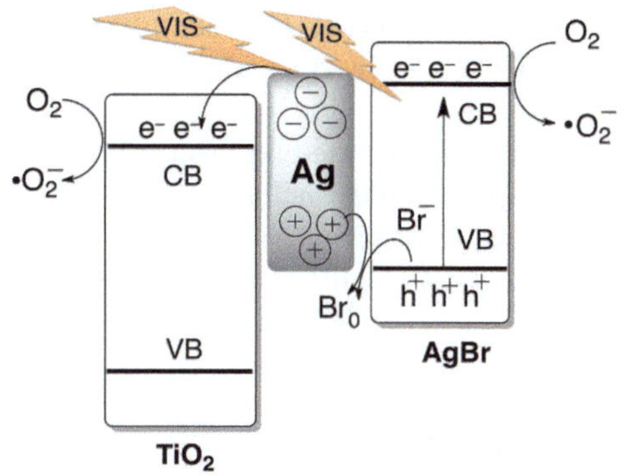

Scheme 3 Schematic diagram for the charge separation in a visible-light irradiated Ag/AgBr/TiO_2 system. Adapted from Ref. [124]

Another similar photocatalyst, AgNP supported on AgBr (Ag@AgBr), catalyzed the decomposition of MO more efficiently than Ag@AgCl under visible light irradiation [116, 119, 120]. Also, reduction of CO_2 to methanol has been observed by plasmon mediated catalysis with Ag@AgBr hexagonal nanoplates [115]. Note AgBr nanoparticles that have shown face-dependent photocatalytic properties for example in the degradation of MO [121]. In this chapter only those semiconductor halides containing AgNP are reviewed. We refer to other papers and the references cited therein for further details on AgX and AgX@semiconductor nanocomposites photocatalytic activity [111, 121].

The photocatalytic activity of Ag@AgI spherical nanoparticles has been recently studied and effectively show antibacterial activity against both E. coli and S. aureus [122]. In all three cases, Ag@AgCl, Ag@AgBr and Ag@AgI plasmon mediated catalysis, the surface of AgX nanoparticles is terminated by X^- ions and therefore, is supposed to be negatively charged [107, 109, 119]. Consequently, AgNP on their surface should polarize the electron distribution such that the regions of its negative charges are far from the Ag/AgX interface, whereas the positive ones are close to it [109]. Then, under visible light illumination, where the SPB of AgNP absorbs, AgNP produce excited electrons and holes [107, 109, 119], thus promoting the photoinduced electron ejection from AgNP into the conduction band of AgX [111, 119]. This is possible because the Fermi energy level of AgX is lower than that of AgNP. Therefore, electrons are transferred from AgNP to AgX until the two systems attain equilibrium and form the new Fermi level [119].

These photogenerated electrons in the conduction band, together with the injected SPR electrons from AgNP, could then initiate the catalytic reaction. Obviously, the interfacial junction between AgNP and AgX could facilitate the charge separation in SPR-excited AgNP and produce long lived charge carriers from the AgNP [119].

As explained above, the holes transferred to the AgX will combine with halide ions X^- ions to form X^0, which will oxidize organic molecules and then reduce back to X^-. In the absence of organic pollutants, X^0 can react with OH^- to form $^{\bullet}OH$ [109, 123]. The electrons would generate superoxide ions and other reactive oxygen species after reacting with O_2 [111].

2.3.1 Heterogeneous Photocatalysts Composed of AgNP Ag/AgX (X = Cl, Br, I) Loaded on Other Supports

Relevant studies based on this type of photocatalysis mechanism have been widely reported; examples include loaded Ag@AgX (X = halide) hybrid structures on TiO_2 [98, 124–126], Al_2O_3 [100, 127], ZnO [128], BiOBr [129], WO_3 [116], Ag_3VO_4 [130], and graphene (also sulphonated graphene [131] and graphene oxide (GO)) [101, 108, 132].

These nanohybrids have proven useful to degrade organic pollutants such as chlorophenols [100, 124, 127], MO [98, 124, 128, 129], rhodamine B [128, 130], acid orange 7 [124], azo-dyes [125], MB [130], volatile organic compounds (VOCs) in the gas phase (benzene and acetone) [126] and also bacteria [116, 125, 133].

The mechanism that Ag@AgX nanoparticles loaded onto semiconductors undergo upon visible light irradiation is depicted in Scheme 3 (Br and TiO_2 have been chosen as example) [98, 124]. The excited electron was transferred to the conduction band of TiO_2, and a hole was subsequently recovered when the bromide ion and bromide radical were formed.

As discussed above, the oxygen molecule can trap the electron injected into the conduction band of the TiO_2, and form superoxide ions, which promoted the photooxidation reaction. Simultaneously, the highly active bromide radical decomposed the organic molecules near the particles [124].

In another recent contribution, AgNP were used to induce visible light photocatalytic reduction of GO leading to reduced graphene oxide (rGO) in the presence of an electron donor [134]. Indeed, AgNP have also been used as a bridge to connect activated GO and $Ag_2Mo_3O_{10}$ or Ag_3PO_4 [101, 135]. This strategy was also described using Ag@AgX nanoparticles [108, 136–138].

Both, GO or rGO can act as an electron acceptors to effectively suppress the charge recombination, resulting in more reactive species [139]. The advantages of their hybridization can be found in most silver halide and phosphate cases [135–138, 140, 141]. When combined with semiconductors, the photoactivity enhancement of semiconductors after hybridization with rGO or GO has been attributed to a 2D network with an electron sink to accept and shuttle electrons photogenerated in the semiconductors [142–144].

Thus, for $Ag_2Mo_3O_{10}$, a reinforced charge transfer and a suppressed recombination of electron hole pairs result from hybridization of Ag@AgX with GO nanosheets, and contribute to the enhanced photocatalytic activity in the degradation of methyl orange [108].

Silver molibdates such as $Ag_2Mo_3O_{10}$ Ws can also generate an electron/hole pair and to reduce Ag^+ ions to Ag^0 particles under illumination, leading to a cluster of silver atoms on an $Ag_2Mo_3O_{10}$ Ws backbone [101]. However, although they show lower photocatalytic activity than silver salt-based photocatalysts, Ag@$Ag_2Mo_3O_{10}$@GO nanohybrids are able to photodegrade rhodamine B an 4-chlorophenol [101].

The proposed mechanism for the activity of Ag^0 is as a solid-state electron mediator that accepts electrons from activated GO and $Ag_2Mo_3O_{10}$ Ws (Scheme 4) [101].

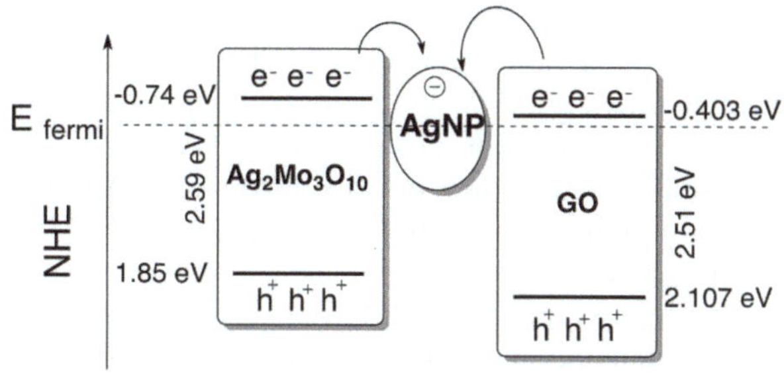

Scheme 4 Photocatalytic process of Ag@$Ag_2Mo_3O_{10}$/activated graphene oxide composite through assistance of solar light. Adapted from reference [101]

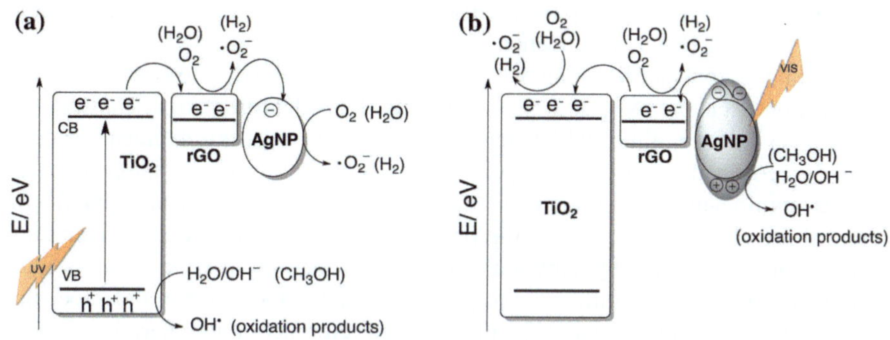

Scheme 5 Proposed mechanism for the photodegradation of RhB and photocatalytic hydrogen production (*in brackets*) by Ag@rGO@TiO_2 under **a** UV part and **b** visible part of the simulated sunlight irradiation. Adapted from from reference [145]

Interestingly, Wang et al. [145] have just reported that Ag@rGO@TiO_2 nanocomposites can act as photocatalysts to generate hydrogen and are also able to photodegrade RhB. The authors claim that in this system the photogenerated electron separation is improved. Different sizes of AgNP were tested. Small sized AgNP (2–5 nm) could store a photoexcited electron that was generated from TiO_2 leading to photocatalysis improvement; whereas, large sized AgNP were able to absorb visible light. Scheme 5 illustrates the two possible pathways for the photocatalytic process in these Ag@rGO@TiO_2 composites.

3 Positive Plasmon Mediated Catalysis

Catalysis is helping to improve chemical conversion, energy production and pollution mitigation. However, most commercial heterogeneous catalytic reactions are being run at relatively high temperatures [90].

Many interesting and useful PMC reactions have been reported using AgNP combined with supports and lower temperatures than their conventional reactions that use only thermal activation [90]. Up to now, water splitting [11], photopolymerization [146–149], isomerization [150], and synthesis of organic compounds [31] have been carried out. Regarding positive PMC, e.g. PMC to make or transform chemicals without destroying them, only CO [90], NH_3 [90] and alcohol oxidations [45, 90, 151], epoxidations [90, 152] and coupling of p-aminobenzenethiol to 4,4'-dimercaptobenzene [112, 153–162] have been explored. Previous examples along this chapter illustrate that most efforts are being devoted to AOPs.

Similar mechanisms for PMC either with AuNPs or AgNP have been described under UV and/or visible light illumination. For AgNP there is an enhanced activity under UV irradiation as compared with visible irradiation, which is sometimes difficult, as explained before, due to the oxidation and/or dissolution of silver. In this chapter, only those examples that use AgNP are mentioned. It is important to remind that: (i) a linear dependence between the temperature and photocatalytic reaction rate at constant light intensity has been observed; (ii) photocatalytic reaction rates on excited MNPs exhibit an intensity dependent transition from the linear to super-linear regime, which indicates an electron-driven chemical reaction [163]. Moreover, this super-linear power law dependence on light intensity takes place at significantly lower intensity than required for super-linear behavior on extended metal surfaces and; (iii) the photocatalytic reaction rate at a given temperature can be obtained by subtracting the rate of the pure thermal process [32]. Readers are strongly recommended to get more insight about linear and super-linear dependences in references [32] and [163] and references cited therein.

After AgNP absorb light, heat is generated and an electron transfer to available adsorbate (reactant) states can occur, forming a negative ion species or a radical. This negative ion can, can either undergo a rapid reaction on the metal surface [90] often referred to as the transient negative ion (TNI) [11, 164–169]; or diffuse to the solution where it could react [45].

This process is very similar to the above-discussed electron injection from an excited AgNP with a nearby semiconductor (i.e. TiO_2, AgX), except that the electron is injected in "adsorbate" states rather than in a semiconductor conduction band [11]. Note that, if the incident light is in the UV range, the formation of charged ion adsorbates could also be initiated by photon-induced interband transition in silver [45].

The interaction of excited plasmons with reactants will be affected by the spatially non-homogeneous distribution of plasmons on the surface of nanoparticles. Therefore, hot spots might play a critical role [11].

Thus, if AgNP are illuminated in the presence of an adsorbate, for example oxygen, the O–O bond is activated at lower temperatures due to a photothermal (electron–phonon driven) elementary step [90]. Scheme 6 contains more detailed information of the formation and subsequent relaxation of the O_2–TNI on the AgNP surface, leading to the dissociation of the adsorbed molecular O_2 [11, 21, 32, 90].

Another mechanism, where a AgNP can transfer electron density to O_2 either directly through chemical interface damping or indirectly through the decay of

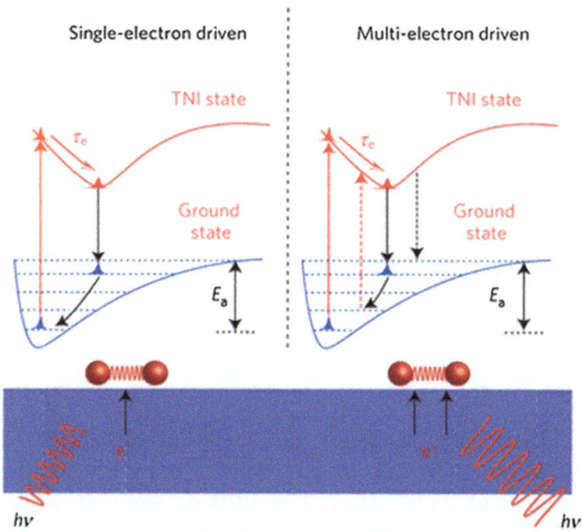

Scheme 6 Proposed molecular mechanism in the linear regime. A single electron excitation deposits vibrational energy into the adsorbate by accelerating the molecule along the TNI potential energy surface (PES) for the lifetime, τ_e. If vibrational energy is not sufficient to overcome the activation barrier, E_a, the adsorbate returns to the thermally equilibrated state. Reprinted with permission from reference [32]. Copyright 2011 Macmillan Publishers Limited

plasmons into energetic electron/hole pairs and subsequent transfer of the e^- to O_2 was proposed by Brus [23]. Irrespective of the mechanism, electrons can transiently populate unoccupied states (orbitals) of adsorbed O_2 and induce electron-mediated O_2 dissociation [23]. Thus, plasmon mediated electron transfer from AgNP to the O_2 $2\pi^*$ orbital forms a transient negative O_2^- species [32, 170, 171].

Factors that influence the capacity of AgNP to drive photocatalytic transformations have been postulated to include:

- (i) temperature-dependent distribution of excited vibrational states, which significantly affects the probability for the plasmon-mediated formation of TNI, and the probability that the TNI will gain sufficient energy to overcome the reaction barrier; and
- (ii) also, much lower activation barriers for chemical transformations on metals as opposed to semiconductors. This allows for lower adsorbate energy required to overcome the activation barrier. Remarkably, Linic et al. suggested that this is the main reason why fairly short-lived TNI on metals (on the order of a few femtoseconds) can induce chemical transformations [11].

AgNP supported on insulators such ZrO_2, Al_2O_3 and SiO_2 or Zeolite Y were found to be highly efficient photocatalysts under visible light for the selective oxidation of benzyl alcohol (Scheme 7, Table 1) [45]. Supports cannot absorb visible light, due to their wide band gaps and there is no electron transfer from the irradiated AgNP to the insulator supports.

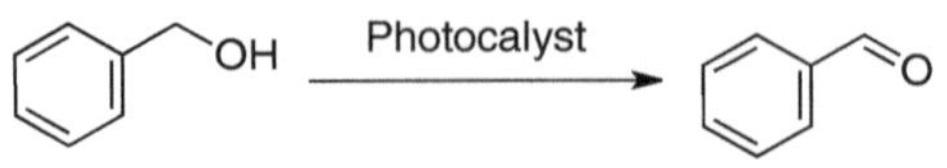

Scheme 7 Oxidation of benzyl alcohol to benzaldehyde

Table 1 Oxidation of benzyl alcohol to benzaldehyde

Photocatalyst	hν	t (h)	Gas	Conversion (%)	Yield (%)
Ag@Zeolite Y	UV	48	Air	11	62
Ag@Zeolite Y	UV	48	O_2	4	100
Ag@TiO_2	Vis	2	O_2	12	12

Ag@TiO_2 with oxygen vacancies gave rise to benzaldehyde under visible light irradiation although better results were reported when using palladium nanoparticles instead of AgNP [51]. The SPB of Ag@TiO_2 cannot be excited with UV. Therefore, upon UV illumination, AgNP on TiO_2 trap the photoexcited electrons from the surface of TiO_2, prolonging the lifetimes of holes and electrons [151]. The holes remaining on the surface of TiO_2 can oxidize absorbed methoxy coming from the thermal dissociation of methanol on TiO_2 to HCHO and HCO. The residual methoxy couples with HCO to form methyl formate [151, 172].

Notably, AgX NPs are also promising materials to reduce CO_2 under visible light irradiation due to the more negative conduction band (CB) edges than $\varphi^{\Theta}(CO_2/CH_3OH)$ and $\varphi^{\Theta}(CO_2/C_2H_5OH)$ (0.38 and 0.085 eV vs. NHE) [111, 112, 115].

Ag@Al_2O_3 nanocomposite has also proven to be an active photocatalyst under visible light irradiation for the selective oxidation (epoxidation) of ethylene to form ethylene oxide (EO), which is a commercially important reaction [32, 90]. The dependence of the photocatalytic rate on the light intensity confirmed that the energetic electrons excited by the LSPR effect of AgNP, directly participate in the photocatalytic process of this oxidation.

Plasmonic excitation of AgNP influences protonation on photoreactions of p-aminobenzenethiol (PATP) leading to its coupling reaction product, 4,4'-dimercaptoazobenzene (DMAB) [153]. In the absence of O_2 or H_2, the plasmon-driven photocatalysis mechanism (hot electron—hole reactions) is the major reaction channel. In the presence of O_2 or H_2, the plasmon-assisted surface catalysis mechanism (activated oxygen/hydrogen reactions) is the major one [154]. Since experimental and theoretical evidence on this surface catalyzed reaction of DMAB produced from PATP, assisted by AgNP in 2010 [173], a series of investigations have been reported to support, or further confirm the conclusions [112, 155–162].

Last but not least, plasmonic mediated water splitting caused by visible light and improvement of photovoltaic devices have attracted attention among scientists [10, 11]. The reader can consult references [10, 11, 38, 174–179] to find more detailed information regarding this topic.

4 Concluding Remarks

MNPs such as AgNP are currently contributing to the development of greener technologies that are based on light driven applications. In this chapter we reviewed several examples of AgNP-support composites that are being used for photocatalysis and plasmon mediated catalytic reactions. Many of them are reusable without losing activity and remain stable under the reaction conditions. Although much effort is being placed in advanced oxidation processes, it is reasonable to assume that significant development will take place in the area of synthetic chemistry promoted by plasmon mediated catalysis. This will allow scientists to utilize the full range of solar energy to activate this new class of photocatalysts for both synthetic and AOPs processes in the near future. This green approach would help to save energy and design more environmentally friendly photocatalytic processes.

Acknowledgments Maria Gonzalez-Bejar thanks the University of Valencia and the Spanish Ministry of Economy and Competitiveness for her Ramón y Cajal contract.

References

1. Galian, R.E., Pérez-Prieto, J.: Catalytic processes activated by light. Energy Environ. Sci. **3**, 1488–1498 (2010)
2. González-Béjar, M., Pérez-Prieto, J., Scaiano, J.C.: Light-driven catalysis. In:Luque, R. (ed.) Green Chemistry, pp. 71–105 Nova Science Publishers, Inc, New York (2012)
3. Klán, P., Wirz, J.: Photochemistry of organic compounds: from concepts to practice. Wiley-Blackwell, Chicester (2009)
4. Hoffmann, M.R., et al.: Environmental applications of semiconductor photocatalysis. Chem. Rev. **95**(1), 69–96 (1995)
5. Khataee, A.R., Fathinia, M.: New and future developments in catalysis: nanoparticle catalysis by surface plasmon, In Sun, S.L. (ed.) Recent advances in photocatalytic processes by nanomaterials, pp. 267–288. Elsevier Science Ltd, United Kingdom (2013)
6. Fujishima, A., Honda, K.: Electrochemical photolysis of water at a semiconductor electrode. Nature **238**, 37–38 (1972)
7. Kudo, A., Miseki, Y.: Heterogeneous photocatalyst materials for water splitting. Chem. Soc. Rev. **38**(1), 253–278 (2009)
8. Zou, Z., et al.: Direct splitting of water under visible light irradiation with an oxide semiconductor photocatalyst. Nature **414**(6864), 625–627 (2001)
9. Borgarello, E., et al.: Photochemical cleavage of water by photocatalysis. Nature **289**(5794), 158–160 (1981)
10. Navarro Yerga, R.M., et al.: Water splitting on semiconductor catalysts under visible-light irradiation. ChemSusChem **2**(6), 471–485, (2009)
11. Linic, S., Christopher, P., Ingram, D.B.: Plasmonic-metal nanostructures for efficient conversion of solar to chemical energy. Nat. Mater. **10**(12), 911–921 (2011)
12. Kafizas, A., et al.: Titanium dioxide and composite metal/metal oxide titania thin films on glass: a comparative study of photocatalytic activity. J. Photochem. Photobiol. A: Chem. **204**(2–3), 183–190 (2009)
13. Giulio, S., et al.: Photo-catalytic coating of polystyrene for household cooling appliances with self cleaning surfaces. J. Appl. Electrochem. **39**(11), 2265–2273 (2009)

14. Bozzi, A., Yuranova, T., Kiwi, J.: Self-cleaning of wool-polyamide and polyester textiles by TiO_2-rutile modification under daylight irradiation at ambient temperature. J. Photochem. Photobiol. A: Chem. **172**(1), 27–34 (2005)
15. Sarina, S., Waclawik, E.R., Zhu, H.: Photocatalysis on supported gold and silver nanoparticles under ultraviolet and visible light irradiation. Green Chem. **15**(7), 1814–1833 (2013)
16. Ueno, K., Misawa, H.: Surface plasmon-enhanced photochemical reactions. J. Photochem. Photobiolo. C: Photochem. Rev. **15**, 31–52 (2013)
17. Wang, P., et al.: Plasmonic photocatalysts: harvesting visible light with noble metal nanoparticles. Phys. Chem. Chem. Phys. **14**(28), 9813–9825 (2012)
18. Eppler, A., et al.: Model catalysts fabricated using electron beam lithography and pulsed laser deposition. J. Phys. Chem. B **101**, 9973–9977 (1997)
19. Goesmann, H., Feldmann, C.: Nanoparticulate functional materials. Angew. Chem. Int. Ed. **49**, 1362–1395 (2010)
20. Mahmoud, M.A., Narayanan, R., El-Sayed, M.A.: Enhancing colloidal metallic nanocatalysis: sharp edges and corners for solid nanoparticles and cage effect for hollow ones. Acc. Chem. Res. **46**(8), 1795–1805 (2013)
21. Linic, S., et al.: Catalytic and photocatalytic transformations on metal nanoparticles with targeted geometric and plasmonic properties. Acc. Chem. Res. **46**(8), 1890–1899 (2013)
22. Nie, S., Emory, S.R.: Probing single molecules and single nanoparticles by surface-enhanced Raman scattering. Science **275**, 1102–1106 (1997)
23. Brus, L.: Noble metal nanocrystals: plasmon electron transfer photochemistry and single-molecule Raman spectroscopy. Acc. Chem. Res. **41**(12), 1742–1749 (2008)
24. Jiang, et al.: Single molecule raman spectroscopy at the junctions of large Ag nanocrystals. J. Phys. Chem. B **107**(37), 9964–9972 (2003)
25. Kelly, K.L., et al.: The optical properties of metal nanoparticles: the influence of size, shape, and dielectric environment. J. Phys. Chem. B **107**(3), 668–677 (2003)
26. Atwater, H.A., Polman, A.: Plasmonics for improved photovoltaic devices. Nat. Mater. **9**(3), 205–213 (2010)
27. Jain, P.K., et al.: Noble metals on the nanoscale: optical and photothermal properties and some applications in imaging, sensing, biology, and medicine. Acc. Chem. Res. **41**(12), 1578–1586 (2008)
28. Stamplecoskie, K.G., Scaiano, J.C.: Light emitting diode irradiation can control the morphology and optical properties of silver nanoparticles. J. Am. Chem. Soc. **132**, 1825–1827 (2010)
29. Chimentão, R.J., et al.: Different morphologies of silver nanoparticles as catalysts for the selective oxidation of styrene in the gas phase. Chem. Commun. pp. 846–847 (2004)
30. Xu, R., et al.: Shape-dependent catalytic activity of silver nanoparticles for the oxidation of styrene. Chem. Asian J. **1**, 888–893 (2006)
31. Sun, M., New and Future Developments in Catalysis: Nanoparticle Catalysis by Surface Plasmon, In: Sun, S.L. (ed.) Catalysis Nanoparticles, pp. 473–487 Elsevier Science Ltd, United Kingdom (2013)
32. Christopher, P., et al.: Singular characteristics and unique chemical bond activation mechanisms of photocatalytic reactions on plasmonic nanostructures. Nat. Mater. **11**(12), 1044–1050 (2012)
33. Corma, A., et al.: Photoinduced electron transfer within zeolite cavities: cis-stilbene isomerization photosensitized by 2,4,6-Triphenylpyrylium cation imprisoned inside zeolite Y. J. Am. Chem. Soc. **116**, 2276–2280 (1994)
34. Corma, A., García, H.: Zeolite-based photocatalysts. Chem. Commun. **13**, 1443–1459 (2004)
35. Protti, S., Fagnoni, M.: The sunny side of chemistry: green synthesis by solar light. Photochem. Photobiol. Sci. **8**, 1499–1516 (2009)
36. Palmisano, G., et al.: Photocatalysis: a promising route for 21st century organic chemistry. Chem. Commun. pp. 3425–3437 (2007)
37. Esser, P., Pohlmann, B., Scharf, H.-D.: The photochemical synthesis of fine chemicals with sunlight. Angew. Chem. Int. Ed. **33**, 2009–2023 (1994)
38. Awazu, K., et al.: A Plasmonic photocatalyst consisting of silver nanoparticles embedded in titanium dioxide. J. Am. Chem. Soc. **130**(5), 1676–1680 (2008)

39. Zhao, Z., Carpenter, M.A.: Support-free bimodal distribution of plasmonically active Ag/AgOx nanoparticle catalysts: attributes and plasmon enhanced surface chemistry. J. Phys. Chem. C **117**(21), 11124–11132 (2013)
40. Petronella, F., et al.: Photocatalytic activity of nanocomposite catalyst films based on nanocrystalline metal/semiconductors. J. Phys. Chem. C **115**(24), 12033–12040 (2011)
41. Hirakawa, T., Kamat, P.V.: Photoinduced electron storage and surface plasmon modulation in Ag@TiO_2 clusters. Langmuir **20**, 5645–5647 (2004)
42. Hirakawa, T., Kamat, P.V.: Charge separation and catalytic activity of Ag@TiO_2 core-shell composite clusters under UV-irradiation. J. Am. Chem. Soc. **127**(11), 3928–3934 (2005)
43. Grabowska, E., et al.: Modification of titanium(IV) dioxide with small silver nanoparticles: application in photocatalysis. J. Phys. Chem. C **117**(4), 1955–1962 (2013)
44. Gong, D., et al.: Silver decorated titanate/titania nanostructures for efficient solar driven photocatalysis. J. Solid State Chem. **189**, 117–122 (2012)
45. Chen, X., et al.: Supported silver nanoparticles as photocatalysts under ultraviolet and visible light irradiation. Green Chem. **12**(3), 414–419 (2010)
46. Chen, X., et al.: Increasing solar absorption for photocatalysis with black hydrogenated titanium dioxide nanocrystals. Science **331**, 746–750 (2011)
47. Chen, X., Mao, S.S.: Titanium dioxide nanomaterials: synthesis, properties, modifications, and applications. Chem. Rev. **107**, 2891–2959 (2007)
48. Zhang, Y., et al.: TiO_2 – graphene nanocomposites for gas-phase photocatalytic degradation of volatile aromatic pollutant: is TiO_2 – graphene truly different from other TiO_2 – carbon composite materials? ACS Nano **4**, 7303–7314 (2010)
49. Hoang, S., et al.: enhancing visible light photo-oxidation of water with TiO_2 nanowire arrays via cotreatment with H_2 and NH_3: synergistic effects between Ti^{3+} and N. J. Am. Chem. Soc. **134**, 3659–3662 (2012)
50. Zhang, N., et al.: Synthesis of M@TiO_2 (M = Au, Pd, Pt) core – shell nanocomposites with tunable photoreactivity. J. Phys. Chem. C **115**, 9136–9145 (2011)
51. Pan, X., Xu, Y.-J.: Defect-mediated growth of noble-metal (Ag, Pt, and Pd) nanoparticles on TiO_2 with oxygen vacancies for photocatalytic redox reactions under visible light. J. Phys. Chem. C **117**(35), 17996–18005 (2013)
52. Xu, Y.-J., Zhuang, Y., Fu, X.: New insight for enhanced photocatalytic activity of TiO_2 by doping carbon nanotubes: a case study on degradation of benzene and methyl orange. J. Phys. Chem. C **114**, 2669–2676 (2010)
53. Kumar, S.G., Devi, L.G.: Review on modified TiO_2 photocatalysis under UV/visible light: selected results and related mechanisms on interfacial charge carrier transfer dynamics. J. Phys. Chem. A **115**, 13211–13241 (2011)
54. Li, R., Kobayashi, H., Guo, J.F., Fan, J.: Visible-Light-Driven Surface Reconstruction of Mesoporous TiO_2: toward visible-light absorption and enhanced photocatalytic activities. Chem. Commun. **47**, 8584–8586 (2011)
55. Zuo, F., et al.: Self-Doped Ti^{3+} enhanced photocatalyst for hydrogen production under visible light. J. Am. Chem. Soc. **132**, 11856–11857 (2010)
56. Zhang, N., et al.: Constructing ternary CdS – Graphene – TiO_2 hybrids on the flatland of graphene oxide with enhanced visible-light photoactivity for selective transformation. J. Phys. Chem. C **116**, 18023–18031 (2012)
57. Zhang, Y., et al.: Engineering the unique 2D mat of graphene to achieve graphene-TiO_2 nanocomposite for photocatalytic selective transformation: what advantage does graphene have over its forebear carbon nanotube? ACS Nano **5**, 7426–7435 (2011)
58. Kim, H.G., et al.: Photocatalytic nanodiodes for visible-light photocatalysis. Angew. Chem. Int. Ed. **44**, 4585–4589 (2005)
59. Ding, C., Zhu, A., Tian, Y.: Functional surface engineering of C-dots for fluorescent biosensing and in vivo bioimaging. Acc. Chem. Res. **47**(1), 20–30 (2014)
60. Cronemeyer, D.C.: Infrared absorption of reduced rutile TiO_2 single crystals. Phys. Rev. **113**, 1222–1226 (1959)
61. Pan, X., et al.: Defective TiO_2 with oxygen vacancies: synthesis, Properties and photocatalytic Applications. Nanoscale **5**, 3601–3614 (2013)

62. Hoang, S., Guo, S., Mullins, C.B.: Coincorporation of N and Ta into TiO_2 nanowires for visible light driven photoelectrochemical water oxidation. J. Phys. Chem. C **116**, 23283–23290 (2012)
63. Hoang, S., et al.: Visible light driven photoelectrochemical water oxidation on nitrogen-modified TiO_2 nanowires. Nano Lett. **12**, 26–32 (2011)
64. Justicia, I., et al.: Designed self-doped titanium oxide thin films for efficient visible-light photocatalysis. Adv. Mater. **14**, 1399–1402 (2002)
65. Pan, X., et al.: Selective oxidation of benzyl alcohol over TiO_2 nanosheets with exposed 001 facets: catalyst deactivation and regeneration. Appl. Catal. A **453**, 181–187 (2013)
66. Ye, M., et al.: High-efficiency photoelectrocatalytic hydrogen generation enabled by palladium quantum dots-sensitized TiO_2 nanotube arrays. J. Am. Chem. Soc. **134**, 15720–15723 (2012)
67. Arabatzis, I.M., et al.: Silver-modified titanium dioxide thin films for efficient photodegradation of methyl orange. Appl. Catal. B- Environ. **42**(2), 187–201 (2003)
68. Ji, Z., et al.: The role of silver nanoparticles on silver modified titanosilicate ETS-10 in visible light photocatalysis. Appl. Catal. B: Environ. **102**, 323–333 (2011)
69. Stathatos, E., et al.: Photocatalytically deposited silver nanoparticles on mesoporous TiO_2 films. Langmuir **16**(5), 2398–2400 (2000)
70. Smirnova, N., et al.: Photoelectrochemical and photocatalytic properties of mesoporous TiO_2 films modified with silver and gold nanoparticles. Surf. Interface Anal. **42**(6–7), 1205–1208 (2010)
71. Liu, S.X., et al.: A mechanism for enhanced photocatalytic activity of silver-loaded titanium dioxide. Catal. Today **93–95**, 877–884 (2004)
72. Messaoud, M., et al.: Photocatalytic generation of silver nanoparticles and application to the antibacterial functionalization of textile fabrics. J. Photochem. Photobiol. A: Chem. **215**, 147–156 (2010)
73. Ingram, D.B., et al.: Predictive model for the design of plasmonic metal/semiconductor composite photocatalysts. ACS Catal. **1**(10), 1441–1447 (2011)
74. Zheng, Z., et al.: Facile in situ synthesis of visible-light plasmonic photocatalysts M@TiO_2 (M = Au, Pt, Ag) and evaluation of their photocatalytic oxidation of benzene to phenol. J. Mater. Chem. **21**, 9079–9087 (2011)
75. Wood, A., Giersig, M., Mulvaney, P.: Fermi level equilibration in quantum dot-metal nanojunctions. J. Phys. Chem. B **105**, 8810–8815 (2001)
76. Georgekutty, R., Seery, M.K., Pillai, S.C.: A highly efficient Ag-ZnO photocatalyst: synthesis, properties, and mechanism. J. Phys. Chem. **112**(35), 13563–13570 (2008)
77. Zheng, Y., et al.: Ag/ZnO heterostructure nanocrystals: synthesis, characterization, and photocatalysis. Inorg. Chem. **46**(17), 6980–6986 (2007)
78. Zheng, Y., et al.: Photocatalytic activity of Ag/ZnO heterostructure nanocatalyst: correlation between structure and property. J. Phys. Chem. C **112**(29), 10773–10777 (2008)
79. Liu, H.R., et al.: Worm-like Ag/ZnO core, shell heterostructural composites: fabrication, characterization, and photocatalysis. J. Phys. Chem. C **116**(30), 16182–16190 (2012)
80. Goto, H., et al.: Quantitative analysis of superoxide ion and hydrogen peroxide produced from molecular oxygen on photoirradiated TiO_2 particles. J. Catal. **225**, 223–229 (2004)
81. Subramanian, V., Wolf, E., Kamat, P.V.: Semiconductor, àímetal composite nanostructures. to what extent do metal nanoparticles improve the photocatalytic activity of TiO_2 films? J. Phys. Chem. B **105**(46), 11439–11446 (2001)
82. Henglein, A., Holzwarth, A., Mulvaney, P.: Fermi level equilibration between colloidal lead and silver particles in aqueous solution. J. Phys. Chem. **96**(22), 8700–8702 (1992)
83. Ohko, Y., et al.: Multicolour photochromism of TiO_2 films loaded with silver nanoparticles. Nat. Mater. **2**(1), 29–31 (2003)
84. Kelly, K.L., Yamashita, K.: The optical properties of metal nanoparticles: the influence of size, shape, and dielectric environment. J. Phys. Chem. B **110**, 7743–7749 (2006)
85. Matsubara, K., et al.: Plasmon resonance-based photoelectrochemical tailoring of spectrum, morphology and orientation of Ag nanoparticles on TiO_2 single crystals. J. Mater. Chem. **19**(31), 5526–5532 (2009)

86. Matsubara, K., et al.: Effects of adsorbed water on plasmon-based dissolution, redeposition and resulting spectral changes of Ag nanoparticles on single-crystalline TiO_2. Phys. Chem. Chem. Phys. **10**(16), 2263–2269 (2008)
87. Wu, T.-S., et al.: Montmorillonite-supported Ag/TiO_2 nanoparticles: an efficient visible-light bacteria photodegradation material. ACS Appl. Mater. Interfaces **2**(2), 544–550 (2010)
88. Li, J., et al.: Ag@Cu_2O Core-shell nanoparticles as visible-light plasmonic photocatalysts. ACS Catal. **3**(1), 47–51 (2013)
89. Stepanov, A.L., Xiao, X., Ren, F., Kavetskyy, T., Osin, Y.N.: Catalytic and biological sensitiviy of TiO_2 and SiO_2 matrices with silver nanoparticles created by ion implantation: a review. Rev. Adv. Mater. Sci. **34**, 107 – 122 (2013)
90. Christopher, P., Xin, H., Linic, S.: Visible-light-enhanced catalytic oxidation reactions on plasmonic silver nanostructures. Nat. Chem. **3**(6), 467–472 (2011)
91. Chen, K.-H., et al.: Ag-Nanoparticle-decorated SiO_2 nanospheres exhibiting remarkable plasmon-mediated photocatalytic properties. J. Phys. Chem. C **116**(35), 19039–19045 (2012)
92. Fuku, K., et al.: The synthesis of size- and color-controlled silver nanoparticles by using microwave heating and their enhanced catalytic activity by localized surface plasmon resonance. Angew. Chem. Int. Ed. **52**(29), 7446–7450 (2013)
93. Burda, C., et al.: Chemistry and properties of nanocrystals of different shapes. Chem. Rev. **105**(4), 1025–1102 (2005)
94. Gunnarsson, L., et al.: Confined plasmons in nanofabricated single silver particle pairs: experimental observations of strong interparticle interactions. J. Phys. Chem. B **109**(3), 1079–1087 (2004)
95. Evanoff, D.D., Chumanov, G.: Synthesis and optical properties of silver nanoparticles and arrays. ChemPhysChem **6**(7), 1221–1231 (2005)
96. Kamat, P.V.: Photophysical, photochemical and photocatalytic aspects of metal nanoparticles. J. Phys. Chem. B **106**(32), 7729–7744 (2002)
97. Mori, K., et al.: Enhancement of the photoinduced oxidation activity of a ruthenium(II) complex anchored on silica-coated silver nanoparticles by localized surface plasmon resonance. Angew. Chem. Int. Ed. **49**(46), 8598–8601 (2010)
98. Yu, J., Dai, G., Huang, B.: Fabrication and characterization of visible-light-driven plasmonic photocatalyst Ag/AgCl/TiO_2 nanotube arrays. J. Phys. Chem. C **113**(37), 16394–16401 (2009)
99. Glaus, S., Calzaferri, G.: The band structures of the silver halides AgF, AgCl, and AgBr: a comparative study. Photochem. Photobiol. Sci. **2**, 398–401 (2003)
100. Hu, C., et al.: Plasmon-induced photodegradation of toxic pollutants with Ag-AgI/Al_2O_3 under visible-light irradiation. J. Am. Chem. Soc. **132**(2), 857–862 (2010)
101. Zhang, K., et al.: Chemically modified graphene oxide-wrapped quasi-micro Ag decorated silver trimolybdate nanowires for photocatalytic applications. J. Phys. Chem. C **117**(45), 24023–24032 (2013)
102. Wang, P., et al.: Synthesis of highly efficient Ag@AgCl plasmonic photocatalysts with various structures. Chem. - Eur. J. **16**(2), 538–544 (2008)
103. Bi, Y., Ye, J.: In situ oxidation synthesis of Ag/AgCl core-shell nanowires and their photocatalytic properties. Chem. Commun. **43**, 6551–6553 (2009)
104. Li, Y., Ding, Y.: Porous AgCl/Ag nanocomposites with enhanced visible light photocatalytic properties. J. Phys. Chem. C **114**(7), 3175–3179 (2010)
105. Lou, Z., et al.: The synthesis of the near-spherical AgCl crystal for visible light photocatalytic applications. Dalton Trans. **40**(16), 4104–4110 (2011)
106. An, C., Peng, S., Sun, Y.: Facile synthesis of sunlight-driven AgCl:Ag plasmonic nanophotocatalyst. Adv. Mater. **22**(23), 2570–2574 (2010)
107. Wang, P., et al.: Ag@AgCl: A highly efficient and stable photocatalyst active under visible light. Angew. Chem. Int. Ed. **47**(41), 7931–7933 (2008)
108. Zhu, M., Chen, P., Liu, M.: Graphene oxide enwrapped Ag/AgX (X = Br, Cl) nanocomposite as a highly efficient visible-light plasmonic photocatalyst. ACS Nano **5**(6), 4529–4536 (2011)

109. Dong, R., et al.: Ecofriendly synthesis and photocatalytic activity of uniform cubic Ag@AgCl plasmonic photocatalyst. J. Phys. Chem. C **117**(1), 213–220 (2013)
110. Han, L., et al.: Facile synthesis of a free-standing Ag@AgCl film for a high performance photocatalyst and photodetector. Chem. Commun. **49**(43), 4953–4955 (2013)
111. Cai, B., et al.: A distinctive red Ag/AgCl photocatalyst with efficient photocatalytic oxidative and reductive activities. J. Mater. Chem. A **2**(15), 5280–5286 (2014)
112. Wu, D.-Y., et al.: Photon-driven charge transfer and photocatalysis of p-aminothiophenol in metal nanogaps: a DFT study of SERS. Chem. Commun. **47**(9), 2520–2522 (2011)
113. Han, L., et al.: Facile solvothermal synthesis of cube-like Ag@AgCl: a highly efficient visible light photocatalyst. Nanoscale **3**(7), 2931–2935 (2011)
114. Tang, Y., et al.: Efficient Ag@AgCl cubic cage photocatalysts profit from ultrafast plasmon-induced electron transfer processes. Adv. Funct. Mater. **23**(23), 2932–2940 (2013)
115. An, C., et al.: Strongly visible-light responsive plasmonic shaped AgX:Ag (X = Cl, Br) nanoparticles for reduction of CO_2 to methanol. Nanoscale **4**(18), 5646–5650 (2012)
116. Wang, P., et al.: Ag/AgBr/$WO_3 \cdot H_2O$: Visible-Light photocatalyst for bacteria destruction. Inorg. Chem. **48**(22), 10697–10702 (2009)
117. Wang, P., et al.: Highly efficient visible light plasmonic photocatalysts Ag@Ag(Cl, Br) and Ag@AgCl-AgI. ChemCatChem **3**(2), 360–364 (2011)
118. Soni, S.S., et al.: Visible-light photocatalysis in titania-based mesoporous thin films. Adv. Mater. **20**(8), 1493–1498 (2008)
119. Jiang, J., Li, H., Zhang, L.: New insight into daylight photocatalysis of AgBr@Ag: synergistic effect between semiconductor photocatalysis and plasmonic photocatalysis. Chem. - Eur. J. **18**(20), 6360–6369 (2012)
120. Wang, H., et al.: Polyhedral AgBr microcrystals with an increased percentage of exposed 111 facets as a highly efficient visible-light photocatalyst. Chem. - Eur. J. **18**(15), 4620–4626 (2012)
121. Wang, H., et al.: Facet-dependent photocatalytic properties of AgBr nanocrystals. Small **8**(18), 2802–2806 (2012)
122. Ghosh, S., et al.: Ag@AgI, Core@Shell structure in agarose matrix as hybrid: synthesis, characterization, and antimicrobial activity. Langmuir **28**(22), 8550–8561 (2012)
123. Lu, M.-C., Chen, J.-N., Chang, C.-P.: Effect of inorganic ions on the oxidation of dichlorvos insecticide with Fenton's reagent. Chemosphere **35**(10), 2285–2293 (1997)
124. Dong, R., et al.: AgBr@Ag/TiO_2 core-shell composite with excellent visible light photocatalytic activity and hydrothermal stability. Catal. Commun. **38**, 16–20 (2013)
125. Hu, C., et al.: Ag/AgBr/TiO_2 Visible light photocatalyst for destruction of azodyes and bacteria. J. Phys. Chem. B **110**(9), 4066–4072 (2006)
126. Zhang, Y., et al.: Nanocomposite of Ag-AgBr-TiO_2 as a photoactive and durable catalyst for degradation of volatile organic compounds in the gas phase. Appl. Catal. B **106**(3–4), 445–452 (2011)
127. Zhou, X., et al.: Plasmon-assisted degradation of toxic pollutants with Ag-AgBr/Al_2O_3 under visible-light irradiation. J. Phys. Chem. C **114**(6), 2746–2750 (2010)
128. Begum, G., Manna, J., Rana, R.K.: Controlled orientation in a bio-inspired assembly of Ag/AgCl/ZnO nanostructures enables enhancement in visible-light-induced photocatalytic performance. Chem. Eur. J. **18**(22), 6847–6853 (2012)
129. Cheng, H., et al.: In situ ion exchange synthesis of the novel Ag/AgBr/BiOBr hybrid with highly efficient decontamination of pollutants. Chem. Commun. **47**(25), 7054–7056 (2011)
130. Zhu, Q., et al.: Facile synthesis of the novel Ag_3VO_4/AgBr/Ag plasmonic photocatalyst with enhanced photocatalytic activity and stability. J. Phys. Chem. C **117**(11), 5894–5900 (2013)
131. Cai, B., et al.: Advanced visible-light-driven photocatalyst upon the incorporation of sulfonated graphene. Nanoscale **5**(5), 1910–1916 (2013)
132. Zhang, H., et al.: Graphene sheets grafted Ag@AgCl hybrid with enhanced plasmonic photocatalytic activity under visible light. Environ. Sci. Technol. **45**(13), 5731–5736 (2011)
133. Wang, X., Lim, T.-T.: Highly efficient and stable Ag-AgBr/TiO_2 composites for destruction of Escherichia coli under visible light irradiation. Water Res. **47**(12), 4148–4158 (2013)

134. Wu, T., et al.: Surface plasmon resonance-induced visible light photocatalytic reduction of graphene oxide: Using Ag nanoparticles as a plasmonic photocatalyst. Nanoscale **3**(5), 2142–2144 (2011)
135. Jiang, B., et al.: In situ fabrication of Ag/Ag_3PO_4/graphene triple heterostructure visible-light photocatalyst through graphene-assisted reduction strategy. ChemCatChem **5**(6), 1359–1367 (2013)
136. Hou, Y., et al.: Ag_3PO_4 Oxygen evolution photocatalyst employing synergistic action of Ag/AgBr nanoparticles and graphene sheets. J. Phys. Chem. C **116**, 20132–20139 (2012)
137. Min, Y., et al.: Self-assembled encapsulation of graphene oxide/Ag@AgCl as a Z-scheme photocatalytic system for pollutant removal. J. Mater. Chem. A **2**(5), 1294–1301 (2014)
138. Luo, G., et al.: Facile fabrication and enhanced photocatalytic performance of Ag/AgCl/rGO heterostructure photocatalyst. ACS Appl. Mater. Interfaces **5**(6), 2161–2168 (2013)
139. Ng, Y.H., et al.: To what extent do graphene scaffolds improve the photovoltaic and photocatalytic response of tio2 nanostructured films? J. Phys. Chem. Lett. **1**(15), 2222–2227 (2010)
140. Liang, Q., et al.: Enhanced photocatalytic activity and structural stability by hybridizing Ag_3PO_4 nanospheres with graphene oxide sheets. Phys. Chem. Chem. Phys. **14**(45), 15657–15665 (2012)
141. Liu, L., Liu, J., Sun, D.D.: Graphene oxide enwrapped Ag_3PO_4 composite: towards a highly efficient and stable visible-light-induced photocatalyst for water purification. Catal. Sci. Technol. **2**(12), 2525–2532 (2013)
142. Xiang, Q., Yu, J., Jaroniec, M.: Graphene-based semiconductor photocatalysts. Chem. Soc. Rev. **41**(2), 782–796 (2012)
143. An, X., Yu, J.C.: Graphene-based photocatalytic composites. RSC Adv. **1**(8), 1426–1434 (2011)
144. Ng, Y.H., Iwase, A., Bell, N.J., Kudo, A., Amal, R.: Semiconductor/reduced graphene oxide nanocomposites derived from photocatalytic reactions. Cat. Today **164**(1), 353–357 (2011)
145. Gao, W., et al.: One-pot synthesis of Ag/r-GO/TiO_2 nanocomposites with high solar absorption and enhanced anti-recombination in photocatalytic applications. Nanoscale **6**(10), 5498–5508 (2014)
146. Stamplecoskie, K.G., et al.: Plasmon-mediated photopolymerization maps plasmon fields for silver nanoparticles. J. Am. Chem. Soc. **133**(24), 9160–9163 (2011)
147. Deeb, C., et al.: Plasmon-based free-radical photopolymerization: effect of diffusion on nanolithography processes. J. Am. Chem. Soc. **133**(27), 10535–10542 (2011)
148. Stamplecoskie, K.G., Fasciani, C., Scaiano, J.C.: Dual-stage lithography from a light-driven, plasmon-assisted process: a hierarchical approach to subwavelength features. Langmuir **28**(30), 10957–10961 (2012)
149. Stamplecoskie, K.G. and J.C. Scaiano, Plasmon mediated polymerization on the surface of silver nanoparticles for advancements in photolithographic patterning. Proc. SPIE, Advances in Resist Materials and Processing Technology XXIX, vol 8325, pp. 832527/1–832527/6 (2012)
150. Hubert, C., et al.: Near-Field Photochemical Imaging of Noble Metal Nanostructures. Nano Lett. **5**(4), 615–619 (2005)
151. Yang, X., et al.: Photocatalytic oxidation of methanol to methyl formate in liquid phase over supported silver catalysts. Catal. Commun. **43**, 192–196 (2014)
152. Zhang, D.-H., et al.: One-pot synthesis of Ag-Fe_3O_4 nanocomposite: a magnetically recyclable and efficient catalyst for epoxidation of styrene. Chem. Commun. **29**, 3414–3416 (2008)
153. Duan, S., et al.: Roles of plasmonic excitation and protonation on photoreactions of p-aminobenzenethiol on Ag nanoparticles. J. Phys. Chem. C **118**(13), 6893–6902 (2014)
154. Zhao, L.-B., et al.: Theoretical study of plasmon-enhanced surface catalytic coupling reactions of aromatic amines and nitro compounds. J. Phys. Chem. Lett. **5**(7), 1259–1266 (2014)
155. Huang, Y.-F., et al.: When the signal is not from the original molecule to be detected: chemical transformation of para-aminothiophenol on Ag during the SERS measurement. J. Am. Chem. Soc. **132**(27), 9244–9246 (2010)
156. Sun, M., et al.: Activated vibrational modes and fermi resonance in tip-enhanced Raman spectroscopy. Phys. Rev. E **87**(2), 020401 (2013)

157. Canpean, V., Iosin, M., Astilean, S.: Disentangling SERS signals from two molecular species: A new evidence for the production of p,p,dimercaptoazobenzene by catalytic coupling reaction of p-aminothiophenol on metallic nanostructures. Chem. Phys. Lett. **500**(4–6), 277–282 (2010)
158. Huang, Y., et al.: Can p, p,-Dimercaptoazobisbenzene be produced from p-Aminothiophenol by surface photochemistry reaction in the junctions of a Ag nanoparticle-molecule-Ag (or Au) Film? J. Phys. Chem. C **114**(42), 18263–18269 (2010)
159. Xu, P., et al.,: Mechanistic understanding of surface plasmon assisted catalysis on a single particle: cyclic redox of 4-aminothiophenol. Sci. Rep.**3** (2013)
160. Sun, M., et al.: The pH-controlled plasmon-assisted surface photocatalysis reaction of 4-Aminothiophenol to p,p´-Dimercaptoazobenzene on Au, Ag, and Cu colloids. J. Phys. Chem. C **115**(19), 9629–9636 (2011)
161. Sun, M., et al.: Remote excitation polarization-dependent surface photochemical reaction by plasmonic waveguide. Plasmonics **6**(4), 681–687 (2011)
162. Sun, M., Hou, Y., Xu, H.: Can information of chemical reaction propagate with plasmonic waveguide and be detected at remote terminal of nanowire? Nanoscale **3**(10), 4114–4116 (2011)
163. Busch, D.G., Ho, W.: Direct observation of the crossover from single to multiple excitations in femtosecond surface photochemistry. Phys. Rev. Lett. **77**, 1338–1341 (1996)
164. Bonn, M., et al.: Phonon- versus electron-mediated desorption and oxidation of CO on Ru(0001). Science **285**(5430), 1042–1045 (1999)
165. Denzler, D.N., et al.: Electronic excitation and dynamic promotion of a surface reaction. Phys. Rev. Lett. **91**(22), 226102 (2003)
166. Buntin, S.A., et al.: Optically driven surface reactions: evidence for the role of hot electrons. Phys. Rev. Lett. **61**(11), 1321–1324 (1988)
167. Olsen, T., Gavnholt, J., Schiøtz, J.: Hot-electron-mediated desorption rates calculated from excited-state potential energy surfaces. Phys. Rev. B **79**(3), 035403 (2009)
168. Olsen, T., Schitøz, J.: Origin of power laws for reactions at metal surfaces mediated by hot electrons. Phys. Rev. Lett. **103**(23), 238301 (2009)
169. Wingreen, N.S., Jacobsen, K.W., Wilkins, J.W.: Inelastic scattering in resonant tunneling. Phys. Rev. B **40**(17), 11834–11850 (1989)
170. Kim, K.H., et al.: Enhanced photoinduced desorption from metal nanoparticles by photoexcitation of confined hot electrons using femtosecond laser pulses. Phys. Rev. Lett. **107**, 047401 (2011)
171. Mulugeta, D., et al.: Size effects in thermal and photochemistry of $(NO)_2$ on Ag nanoparticles. Phys. Rev. Lett. **101**, 146103 (2008)
172. Phillips, K.R., et al.: Sequential photo-oxidation of methanol to methyl formate on TiO2(110). J. Am. Chem. Soc. **135**, 574–577 (2013)
173. Fang, Y., et al.: Ascertaining p, p, Dimercaptoazobenzene produced from p-Aminothiophenol by selective catalytic coupling reaction on silver nanoparticles. Langmuir **26**(11), 7737–7746 (2010)
174. Ingram, D.B., Linic, S.: Water splitting on composite plasmonic-metal/semiconductor photoelectrodes: evidence for selective plasmon-induced formation of charge carriers near the semiconductor surface. J. Am. Chem. Soc. **133**(14), 5202–5205 (2011)
175. Sun, T., et al.: High photocatalytic activity of hydrogen production from water over Fe doped and Ag deposited anatase TiO_2 catalyst synthesized by solvothermal method. Chem. Eng. J. **228**, 896–906 (2013)
176. Wei, Y., et al.: Polydopamine-assisted decoration of ZnO nanorods with Ag nanoparticles: an improved photoelectrochemical anode. J. Mater. Chem. A **1**(16), 5045–5052 (2013)
177. Solarska, R., Krolikowska, A., Augustynski, J.: Silver nanoparticle induced photocurrent enhancement at WO_3 photoanodes. Angew. Chem. Int. Ed. **49**(43), 7980–7983 (2010)
178. Chuang, H.-Y., Chen, D.-H.: Fabrication and photoelectrochemical study of Ag@TiO_2 nanoparticle thin film electrode. Int. J. Hydrogen Energy **36**(16), 9487–9495 (2011)
179. Wei, Y., et al.,: Enhanced photoelectrochemical water-splitting effect with a bent ZnO nanorod photoanode decorated with Ag nanoparticles. Nanotechnology, **23**(23), 235401/1–235401/8 (2012)

Biomedical Uses of Silver Nanoparticles: From Roman Wine Cups to Biomedical Devices

Hasitha de Alwis Weerasekera, May Griffith and Emilio I. Alarcon

Abstract This chapter reviews the state of the art applications of silver nanoparticles in the biomedical sciences and health care, and aims to provide the reader with a general, but concise picture of the basic concepts that need to be considered when using silver nanoparticles (AgNP) in biological systems. We begin with a review of basic concepts on nanomaterials for biomedical applications. In the second part, we discuss a long-standing controversy regarding the *risks* versus *benefits* for the use of AgNP in the biomedical sciences. In the last section of this chapter, we present a discussion of future challenges regarding the use of AgNP in medicine.

An erratum to this chapter is available at DOI 10.1007/978-3-319-11262-6_7

H. de Alwis Weerasekera
Department of Chemistry and Centre for Catalysis Research and Innovation, University of Ottawa, Ottawa, Canada

M. Griffith (✉)
Department of Clinical and Experimental Medicine, Integrative Regenerative Medicine Centre, Linköping University, Linköping, Sweden
e-mail: may.griffith@liu.se

E.I. Alarcon (✉)
Bio-nanomaterials Chemistry and Engineering Laboratory, Division of Cardiac Surgery, University of Ottawa Heart Institute, 40 Ruskin Street Rm H5229, Ottawa K1Y 4W7, Canada
e-mail: ealarcon@ottawaheart.ca

E.I. Alarcon
Centre for Catalysis Research and Innovation, University of Ottawa, Ottawa, Canada

E.I. Alarcon et al. (eds.), *Silver Nanoparticle Applications*, Engineering Materials,
DOI 10.1007/978-3-319-11262-6_5

1 Introduction

Silver has been present in human civilization since ancient times; for example, the "*Cantharus*" was a two-handle elongated Roman wine cup made of pure silver [1, 2]. Silver-made containers used for water storage were noted to not have any growth of microorganisms—possibly the earliest recorded observation of the antimicrobial properties of silver, before nanomaterials revolution in our present society [3].

Long before the development of nanomaterials in the 20th century, silver played a pivotal role in the medical sciences. Two of the main biomedical uses for ionic silver were as prophylactic eye drops for newborns against gonorrhea and other bacterial/fungal infections [4], and in the so called 'silver-baths', that were common practice in burn units to prevent biofilm formation [5]. Colloidal silver was hailed as the "magic bullet" when Erlich began treating syphilis with silver-salvarsan in 1910, and this became the treatment of choice [6].

It should be noted that in some countries silver nitrate eye drops are still used; although silver baths are now part of the history of modern medicine due to their elevated toxicity [2, 7, 8]. As a consequence of such elevated toxicity, medical burn units replaced the "silver-baths" for the topical application, in the form a 1.0 % cream, with another silver "form"; silver sulfadiazine [8]. The use of this "new" form of silver had been justified due to its apparent lower toxicity and synergetic antibacterial activity between the silver ions and the sulfadiazine moiety. However, studies have raised concerns regarding the cell toxicity of silver sulfadiazine in primary skin cells [9] and the immune suppressing ability of this compound [10]. Colloidal silver is no longer used to treat syphilis, but many individuals consume colloidal as a health supplement, despite concerns over adverse effects such as agyria [11].

Although the exact mechanism for the antibacterial activity of silver is not clear (see Chap. "Anti-microbiological and Anti-infective Activities of Silver"), its use in the form of a nano-sized material offers hope in overcoming the silver toxicity, since the available ionic silver concentration will be dramatically reduced for a fixed total silver concentration; see below.

Available silver concentration calculation

Let us consider an illustrative example for the calculation of the available silver concentration for 3.5 nm diameter spherical AgNP prepared using 0.2×10^{-3} mol/L ionic silver concentration. AgNP concentration and surface density/concentration can be calculated in a fashion similar to that described for spherical gold nanoparticles (AuNPs) [12]. In this calculation, we have assumed complete Ag^{+} reduction to form 3.5 nm spherical AgNP. Thus, for a large spherical metal cluster, it can be assumed that the volume of the cluster (V_{NP}) is equal to N times the volume of each individual atom (V_{Atom}) as follows:

$$V_{NP} = N \times V_{Atom} \quad (1)$$

Using nanoparticle diameter and volume, the total number of atoms contained in a single nanoparticle can be calculated as follows:

$$N = \left(\frac{D_{NP}}{d_A}\right)^3 \quad (2)$$

In (2), D_{NP} and d_A are the nanoparticle and silver atomic diameters, respectively. We have used in our calculations $D_{NP} = 3.5$ nm and $d_A = 0.259$ nm. This corresponds to an average of 2,468 atoms per nanoparticle (N). From this, AgNP concentration can be calculated using Eq. 3:

$$[AgNP] = \frac{[Ag]}{N} \quad (3)$$

Here, Ag concentration is in all cases considered equal to 0.2×10^{-3} mol/L. This calculation renders a AgNP concentration of ≈80 nM (80×10^{-9} mol/L). From this value, and assuming that the available silver concentration will be exclusively found on the surface of the nanoparticle—which for practical purposes will have a thickness of at least ≈0.26 nm (the diameter of a single Ag atom)—the percentage of silver atoms in one AgNP can be approximately calculated, as follows:

$$\%Ag - surface = \frac{d^3 - (d-1)^3}{d^3} \times 100 \quad (4)$$

In Eq. 4, d corresponds to the nanoparticle diameter (3.5 nm) and l is the thickness of the outer layer (0.26 nm). Thus, for 3.5 nm diameter spherical silver nanoparticles, around 21 % of the atoms are contained on the nanoparticle surface, which corresponds to 16.5 nM silver. Table 1 details the number of silver atoms per nanoparticle, nanoparticle number per liter, nanoparticle concentrations, and available surface silver concentration, for spherical silver nanoparticles with different diameters.

Together with the reduced silver concentration in AgNP, the role of the protecting agent on the nanoparticle surface also plays a critical role in the nanomaterial biocompatibility, since this capping agent controls the interaction of the nanoparticle with living organisms; see below. *Note that the argument related to the reduced silver concentration for nanoparticles is solely and exclusively valid under conditions in which nanoparticle stability and minimal surface oxidation are guaranteed.* This will be further discussed in the following section.

Table 1 Number of silver atoms per nanoparticle, nanoparticle density (#/L), nanoparticle concentrations (nM), and available surface silver concentration (nM) for different sizes of spherical monodisperse silver nanoparticles

Diameter/nm	Ag atoms/ nanoparticle ($\times 10^3$)	AgNP concentration (nM)	AgNP density ($\times 10^{10}$/L)	Available AgNP on surface (nM)
2.5	0.9	220	19	62 (28)[a]
5.0	7.2	28	2.4	4.1 (15)
10	58	3.5	0.30	0.26 (7.5)
25	900	0.22	0.019	0.0068 (3.1)
50	7,195	0.028	0.0024	0.0004 (1.6)
100	57,557	0.0035	0.00030	0.00003 (0.8)

[a]Numbers in brackets correspond to the percentage of atoms that are on the surface of the nanoparticle. Note that we have limited our consideration of surface atoms to those contained within an outer layer of 0.26 nm thickness

2 Revisiting Silver Nanoparticles Toxicity

Ionic silver is well known for its toxic effects on primary cells. However, when compared to AgNP at the same total silver concentration, one should suspect to see a reduced toxicity since the available silver concentration will be considerably lower, as shown in Table 1. Moreover, the greatly increasing number of peer-review publications on AgNP, see Chap. "Synthetic Routes for the Preparation of Silver Nanoparticles: A Mechanistic Perspective", present conflicting evidence regarding the biocompatibility and toxic impact of AgNP in living organisms. Thus, drawing any conclusion regarding the application of nano silver in the biomedical sciences is certainly problematic. In the following, we provide the reader with some practical guidelines to be considered when determining the biocompatibility of AgNP that include:

i. *Nanoparticle size and polydispersity*
ii. *Nature of the capping/protecting agent*
iii. *Nanoparticle stability in the chosen cell culture media or biologically-relevant liquids*
iv. *Experimental differences in the cell viability assays* e.g. *cell density, cytotoxicity assay protocol, and incubation time*

In the next paragraphs, we will briefly discuss the importance that points i–iv have on the evaluation of the toxicity of AgNP.

2.1 Nanoparticle Size and Polydispersity

Numerous reports indicate that nanoparticle size controls the cell uptake mechanism [13, 14]. Those mechanisms can be then divided into two major groups; simple and facilitated diffusion. However, under conditions where close contact between nanoparticles and cells has been achieved, i.e. longer incubation times, a simple calculation using the number of nanoparticles/L listed in Table 1, at a fixed cell concentration (5.0×10^4 cells/mL), renders nanoparticle/cell ratios of 480 and 0.006 for 5 and 100 nm AgNP, respectively. With those numbers and assuming that only one nanoparticle is required to induce complete toxicity, irrespective of the uptake mechanism, for the 5.0 nm AgNP a 100 % mortality should be observed, while for the 100 nm AgNP less than 0.6 % of the population should show signs of toxicity. In this scenario *for the* 100 nm *AgNP lack of toxicity would be observed only if nanoparticle degradation* (i.e. *release of ionic Ag*) *during the incubation is minimal and does not contribute to the overall toxicity.*

Since reduced polydispersity (i.e. tighter size distribution of nanoparticle size), allows for more reliable data interpretation for nanoparticle behavior, particularly when studying nanoparticle-cell interactions, the synthesis of nanoparticles with narrower size distributions is required. However, nanoparticle formation and growth involves multiple steps (see Chap. "Synthetic Routes for the Preparation of Silver Nanoparticles: A Mechanistic Perspective"), resulting in the formation of nanoparticles within a range of sizes. Figure 1 left shows a representative plot for the size distribution of silver nanoparticles protected with a thiol-modified variety of the antimicrobial peptide LL-37 (AgNP@LL37) [9]. This histogram was obtained from measuring 400 individual nanoparticles from TEM images.

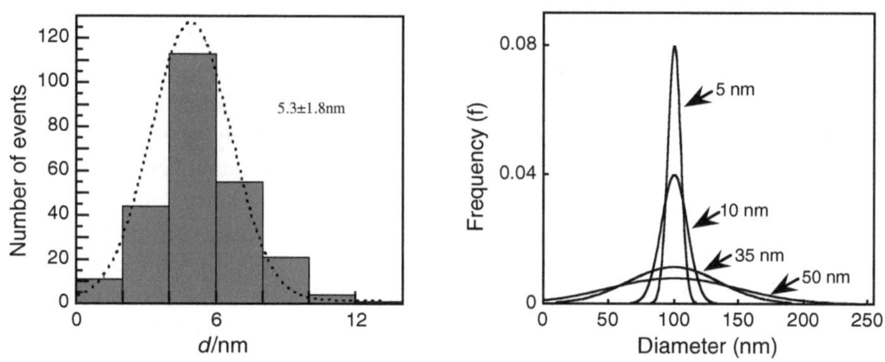

Fig. 1 *Left* Size distribution histogram for AgNP@LL37 nanoparticles obtained from size counting of TEM images for 400 individual nanoparticles. The *dashed line* corresponds to the best Gaussian data fit obtained in Kaleida Graph® 4.5.1. Figure adapted from reference [9]. *Right* Effect sample polydispersity on the size distribution of spherical nanoparticles calculated using a normal distribution Gaussian equation. The numbers in the plot correspond to the standard deviation (SD) used in the curve simulation. In all cases the area under the *curve* was normalized to the unit

As mentioned, nanoparticles are formed within a range of sizes; thus, depending on the size dispersion, they are grouped in two main categories; monodisperse and polydisperse. In this sense, polydispersity index (PDI) accounts for the sample size of colloidal nanoparticles in a fluid solution and it can be understood as a measurement of the size-histogram distribution broadness that can be defined as:

$$PDI = \left(\frac{\sigma}{\chi}\right)^2 \tag{5}$$

In this equation σ and χ correspond to the standard deviation and mean size diameter, respectively. Let us consider samples A and B, both with mean size distributions of 100 nm, but with standard deviations 10 and 35 nm. Using those numbers in Eq. 5 renders PDI values of 0.01 and 0.123 for A and B, respectively; see Fig. 1 right. In general terms, colloidal solutions of nanoparticles with PDI $<$ 0.05 are considered as monodisperse, while values between 0.05 and 0.10 correspond to samples with good monodispersity, and those with PDI $>$ 0.1 are considered polydisperse. Samples with PDI $>$ 0.7 correspond to highly polydisperse samples, where multiple populations of nanoparticles co-exist.

The concept of polydispersity also introduces an additional variable in the number of nanoparticles available at a fixed concentration of starting material, as depicted in Fig. 1 right. This figure shows the impact of polydispersity on the relative abundance of spherical nanoparticles of different diameters in the solution. In this model, the number of available nanoparticles with the same diameter per volume unit will not be the same as that expected if a 100 % of the population had the same size distribution. Thus, these differences in distributions will markedly impact the interpretation for the interaction between nanomaterials and living organisms if samples with wide polydispersity are employed. Table 2 shows a summary of selected properties for different spherical silver nanoparticles employed in recently published articles. The collection of papers shown in this table is relatively small in comparison to what is shown in Chap. "Synthetic Routes for the Preparation of Silver Nanoparticles: A Mechanistic Perspective" regarding the number of peer-reviewed articles published per year on silver nanoparticles. For pedagogical reasons, we have purposely selected papers that have been published during the last 4 years and that meet the quality criteria for complete material characterization and biocompatibility testing. Figure 2 shows the size distribution histogram for the data collected in Table 2. It can be seen that the nanoparticle sizes ranged from a couple of nanometers to up to 100 nm, with nanoparticles smaller than 40 nm being the predominant size found. Statistical analysis of this population rendered a mean number of $\approx$35 nm for nanoparticle diameter with a standard deviation of 27 nm, which indicates a highly polydisperse population. Note that nanoparticle sizes cannot be directly compared when using two different techniques to estimate their sizes. Thus, usually electron microscopy techniques render values more precise for the metal core of the nanoparticle, while dynamic light scattering methodologies, for example, deal with the estimation of the hydrodynamic radii of nanoparticles; see Chap. "Synthetic Routes for the Preparation of Silver Nanoparticles: A Mechanistic Perspective". In the latest the role of the protecting agent and medium ionic strength will play a critical role in the measured hydrodynamic size.

Table 2 Physical properties and cell toxicity conditions for selected AgNP recently reported in the literature

Author	Size (nm) [Measurements]	Capping agent	Initial [Ag]/(M)	Zeta potential (mV)	Particle stability test	Cell line(s)	In situ cytotoxic assay	Incubation time (h)	Cell density (cells/cm^2)[15]	LD_{50} (μg/mL)
Vignoni et al. [9]	19.0 ± 1.0 [DLS]	LL37 peptide	0.0002	+37.0 ± 0.8	AgNP in buffer and cell media ~24 h [UV-Vis]	Human dermal fibroblast	1. MTS colorimetric assay[3] 2. Live/dead staining[4]	14 (1)	4,808 (1), 3,500 (2)	No significant cytotoxicity observed at tested concentrations
Alarcon et al. [15]	3.00 [TEM]	HSA	0.0002	+41.0 ± 3.0	AgNP in buffer and cell media ~18 h [UV-Vis]	Human dermal fibroblast	MTS colorimetric assay[3]	6, 24, 48	4,808	No significant cytotoxicity observed at tested concentrations and time points
Albers et al. [16]	50.0[1, 2]	α- MEM [AgNP resuspension]	NP	NP	NP	(a) Primary osteoblasts (b) Primary osteoclasts	1. XTT colorimetric assay[3]—(a, b) 2. Alkaline phosphatase assay[5] (a) 3. Tartrate resistant acidic phosphatase assay[6] (b)	72	1,923 (a), 28,846 (b)	(a). 194.3 (1), 146.1 (2) (b). 335.0 (1), 193.5 (3)
Ashraf et al. [17]	10.0–15.0 [TEM]	Casein (milk protein)	0.0010	−25.0, −25.0	Intracellular AgNP [Confocal laser scanning microscopy]	Mouse embryonic fibroblast—NIH/3T3	Resazurin based fluorescent assay[7]	24	14, 423	0.314 ± 0.075

(continued)

Table 2 (continued)

Author	Size (nm) [Measurements]	Capping agent	Initial [Ag]/(M)	Zeta potential (mV)	Particle stability test	Cell line(s)	In situ cytotoxic assay	Incubation time (h)	Cell density (cells/ cm^2)[15]	LD_{50} (μg/mL)
Grosse et al. [18]	10.0[1,2], 50.0[1,2], 100.0[1,2]	Citrate [AgNP resuspension]	NP	NP	NP	Rat brain endothelial-RBE4	1. Neutral red assay[8,9] 2. LDH assay[9]	24 (1)	30,000 (1)	In-situ cytotoxicity test (1). 1-10 (10 nm), 10-25 (50 nm), >25 (100 nm)
Kermanizadeh et al. [19]	17.5[1] [TEM]	Polyoxylaurat Tween-20	NP	NP	AgNP in buffer and cell media [DLS]	Human hepatoblastoma-C3A	1. WST-1 colorimetric assay[3] 2. AlamarBlue fluorescent assay[7]	24	9,615	3.906–7.810 (1), 7.810–15.62 (2)
Simpson et al. [20]	4.4 ± 1.0 [TEM]	Type-I collagen @ rose bengal	0.0002	30.0	NP	Human dermal fibroblast	Light microscopy imaging	0, 48, 72, 96	5,000	No significant cytotoxicity observed at tested concentrations and time points
Suman et al. [21]	30.0–55.0 [TEM]	*Morinda citrifia* root extract	0.0010	NP	NP	HeLa	MTT colorimetric assay[3]	24	2,403	~100.0–1000
Alarcon et al. [22]	3.50 ± 0.04 [SEM]	(i) Type-I collagen (ii) α-Poly-L-Lysine	0.0002	+30.0 (i), +70.0 (ii)	AgNP in buffer and cell media 68 (i) and 39 days (ii) [UV-Vis]	(a) Human dermal fibroblast (b) Human epidermal keratinocyte	1. MTS colorimetric assay[3] 2. Live/dead staining[4]	14 (1)	4,808 (1)	In-situ cytotoxicity test (1). (i) 0.0018 (a) (ii) 0.0003 (a), 0.0001(b)

(continued)

Table 2 (continued)

Author	Size (nm) [Measurements]	Capping agent	Initial [Ag]/(M)	Zeta potential (mV)	Particle stability test	Cell line(s)	In situ cytotoxic assay	Incubation time (h)	Cell density (cells/ cm^2)[15]	LD_{50} (µg/mL)
Asare et al. [23]	20.0[1] [DLS~154.6]	BSA [AgNP resuspension]	NP	NP	AgNP in buffers [DLS]	(a) Human testicular embryonic carcinoma—NT2 (b) Primary mouse testicular (WT and KO)	1. MTT colorimetric assay[3] 2. Propidium Iodide/ Hoechst staining[9]	24, 48, 72 (1)	NR	In-situ cytotoxicity test (1). (a). ~100.0 (24h), >100.0 (48 h); (b-WT). 10.00–15.00 (24 h), ~10 (48 h); (b-KO). ~10.00 (24 h), 10.00 (48 h)
Haase et al. [24]	20.0 ($AgNP_{20}$), 40.0 ($AgNP_{40}$) [TEM]	Peptide (CKK)	0.2000	NP	AgNP suspension in buffer and cell media [DLS, SAXS]	Mixed primary cells–Wistar rat neurons and astrocytes (Cells were cultured for 7, 14 & 21 days)	1. WST-1 colorimetric assay[3] 2. LDH assay[9]	24	NR	In-situ cytotoxicity test (1). ($AgNP_{20}$)—~10.00 (7d), ~30.00 (14d), ~20.00 (21d). ($AgNP_{40}$)—~10.00 (7d), 5.000–10.00 (14d), ~10.00 (21d). In-situ cytotoxicity test (2) for 14 d cultures. ($AgNP_{20}$)—0–5.000, ($AgNP_{40}$)—30.00–50.00

(continued)

Table 2 (continued)

Author	Size (nm) [Measurements]	Capping agent	Initial [Ag]/(M)	Zeta potential (mV)	Particle stability test	Cell line(s)	In situ cytotoxic assay	Incubation time (h)	Cell density (cells/cm^2)[15]	LD_{50} (μg/mL)
Hayashi et al. [25]	83.0 ± 22.0[1] [TEM]	BSA [AgNP resuspension]	NP	−10.0–−14.0	Dispersion of AgNP in cell media [DLS, TEM], Intracellular AgNP [TEM]	(a) Earthworm Coelomocyte (b) Human acute monocytic leukemia (THP-1) (c) Differentiating human acute monocytic leukemia (diff. THP-1)	WST-8 colorimetric assay[3]	24	NR	(a) 5.020 ± 1.030 (b) Not tested for the concentration range (c) 7.340 ± 3.800
Justin Packia Jacob et al. [26]	17.6–41.0 [SEM]	*Piper longum* leaf extract	0.0010	NP	NP	Human epithelial (HEp-2)	MTT colorimetric assay[3]	24	500,000	31.25
Lu et al. [27]	12.0 [TEM], 154.2 [DLS]	Egg white extract (albumen)	0.0100	NP	NP	Mouse fibroblast	WST-8 colorimetric assay[3]	24	4,808	No significant toxicity observed at tested concentrations
Mukherjee et al. [28]	28.4[1](dH_2O), 65.4[1] (HeLa med.), 68.6[1] (HeCaT med.) [DLS]	HeCaT media, HeLa media [AgNP resuspension]	NP	−33.33, −20.67, −15.83	Dispersion of AgNP in cell media [DLS]	(a) Human keratinocyte (HeCaT) (b) Epithelial adenocarcinorma (HeLa)	1. AlamarBlue fluorescent assay[7] 2. Neutral red assay[8, 9] 3. Coomassie blue assay[10] 4. MTT colorimetric assay[3] 5. Apoptosis/necrosis assay	24 (1–5), 48 (1–5), 72 (1–5), 96 (1–3)	9,615 (1–4)	Data shown are for 24 h incubations. (a). 181.0(1), 192.5(2), 80.60(3), 51.80(4) (b). 88.60(1), 180.6(2), 80.90(3), 56.40(4)

(continued)

Table 2 (continued)

Author	Size (nm) [Measurements]	Capping agent	Initial [Ag]/(M)	Zeta potential (mV)	Particle stability test	Cell line(s)	In situ cytotoxic assay	Incubation time (h)	Cell density (cells/ cm^2)[15]	LD_{50} (μg/mL)
Suresh et al. [29]	4.5 ± 1.5 (i), 4.0 ± 1.5 (ii), 9.0 ± 2.0 (iii), 4.0 ± 1.0 (iv) [TEM]	(i) Poly (diallyldimethyl-ammonium) chloride (ii) Biogenic (iii) Uncoated (iv) Oleate	0.0010	+45.0 ± 3.1 (i), −12.0 ± 2.0 (ii), −42.5 ± 5.2 (iii), −45.8 ± 4.4 (iv)	AgNP in buffer and cell media [DLS]	(a) Murine peritoneal macrophage (RAW 264.7) (b) Murine lung epithelial (C-10)	1. MTT colorimetric assay[3] 2. Confocal microscopy imaging	4, 12	4,808	In-situ cytotoxicity test (1) at 4 h incubation. (a). 0.100 (i), 0.125(ii), 4.900 (iii), 1.100 (iv) (b). 0.450 (i), 0.700 (ii), 6.300 (iii), 1.600 (iv)
Bouwmeester et al. [30]	20.3 ± 1.9[1], 34.4 ± 3.4[1], 61.2 ± 5.3[1] [TEM]	DMEM [AgNP resuspension]	NP	NP	Dispersion of AgNP in cell media ~0–24 h [TEM, NTA]	Human colon colorectal adenocarcinoma (Caco2)	WST-1 colorimetric assay[3]	24	3,846	No significant toxicity observed at tested concentrations
Comfort et al. [31]	9.15 ± 1.5[1] [TEM]	Uncoated	NP	−40.9 ± 1.9	Dispersion of AgNP in buffer and cell media [DLS], Intracellular AgNP [TEM]	Human epithelial (A-431)	MTS colorimetric assay[3]	24	7,212	~25.00
Foldbjerg et al. [32]	69.0 ± 3.0[1] [TEM]	PVP	NP	−21.8 (stock), −11.6 (media)	Dispersion of AgNP in cell media ~24 h [DLS]	Human alveolar cell line (A549)	1. MTT colorimetric assay[3] 2. Apoptosis/ necrosis assay	24	NR	~ 12.50 (1), ~ 5.000–10.00 (2)

(continued)

Table 2 (continued)

Author	Size (nm) [Measurements]	Capping agent	Initial [Ag]/(M)	Zeta potential (mV)	Particle stability test	Cell line(s)	In situ cytotoxic assay	Incubation time (h)	Cell density (cells/cm^2)[15]	LD_{50} (µg/mL)
Haase et al. [33]	20.0[1] ($AgNP_{20}$), 40.0[1] ($AgNP_{40}$) [TEM]	Peptide (CKK)	0.2000	NP	Intracellular AgNP [TEM, confocal Raman microscopy], AgNP in aqueous suspension [DLS, SAXS]	THP-1 derived human macrophage	1. WST-1 colorimetric assay[3] 2. LDH assay[9]	24 (1), 48 (1)	9,615 (1)	In-situ cytotoxicity tests (1). ($AgNP_{20}$)–110.0 (24 h), 18.00 (48 h) ($AgNP_{40}$)–140.0 (24 h), 30.00 (48 h)
Hackenberg et al. [34]	46.0 ± 21.0[1] [TEM]	BSA [AgNP resuspension]	NP	−13.6	Intracellular AgNP [TEM], Dispersion of AgNP in cell media [DLS]	Human mesenchymal stem	1. Trypan blue exclusion assay[9] 2. Live/dead staining[11]	1(1), 3(1), 24(1)	NR	LD_{50} levels are not reported
Park et al. [35]	20.3 ± 1.9[1], 79.8 ± 5.1[1], [TEM]	Uncoated	NP	−47.1 ± 1.9, −49.7 ± 1.8	DLS	(a) Murine peritoneal macrophage (RAW 264.7) (b) Mouse fibroblast (L929)	1. WST-1 colorimetric assay[3] 2. LDH assay[9]	24 (1), 240 (1)	19,230	LD_{50} levels are not reported

(continued)

Table 2 (continued)

Author	Size (nm) [Measurements]	Capping agent	Initial [Ag]/(M)	Zeta potential (mV)	Particle stability test	Cell line(s)	In situ cytotoxic assay	Incubation time (h)	Cell density (cells/ cm^2)[15]	LD_{50} (μg/mL)
Stevanović et al. [36]	69.2 ± 6.0 (i), 47.5 ± 5.0 (ii), 44.9 ± 5.0 (iii)	(i) Uncoated, (ii) Poly-α, γ, L-glutamic acid (0.05 wt%) (iii) Poly-α, γ, L-glutamic acid (0.1 wt%)	0.0058	−38.5 ± 11.5 (i), −49.1 ± 19.0 (ii), −43.7 ± 12.0 (iii)	~3 months [UV-Vis]	Human liver carcinoma (HepG2)	MTT colorimetric assay[3]	24	3,846	LD_{50} levels are not reported
Stoehr et al. [37]	30.0 [TEM]	PVP	0.1000	NP	AgNP in buffer and cell media [DLS]	Human alveolar epithelial carcinoma	1. Resazurin based colorimetric assay[7] 2. LDH assay[9]	24, 48	1,923	No significant toxicity observed at tested concentrations
Braydich-Stolle et al. [38]	15.0[1], 25.0[1], 80.0[1] (i) 10.0[1], 25.0-30.0[1], 80.0[1] (ii) [DLS, TEM]	(i) Hydrocarbon, (ii) Poly-saccharide	NP	NP	~24 h post exposure intracellular AgNP [SEM, TEM, DLS]	Mouse spermatogonial stem cells (C18-4)	1. MTS colorimetric assay[3] 2. Apoptosis/ Necrosis assay	24 (1), 48 (2)	4,808 (1), 4,250 (2)	LD_{50} levels are not reported
Eom and Choi [39]	28.0–35.0[1] [DLS]	THF [AgNP resuspension]	NP	30.0	Dispersion of AgNP in buffer and cell media [TEM, DLS]	Human lymphoma (Jurkat T)	1. Trypan blue exclusion assay[9] 2. Cell cycle analysis[12] 3. Live/ dead staining[13]	24 (1)	NR	In-situ cytotoxicity test (1). ~0.200–0.500
Martinez-Gutierrez et al. [40]	20.0–25.0 [TEM]	Gallic acid	0.0010	NR	NP	Human acute monocytic leukemia (THP-1)	Trypan blue exclusion assay[9]	24	28,846	10.00 ± 3.40

(continued)

Table 2 (continued)

Author	Size (nm) [Measurements]	Capping agent	Initial [Ag]/(M)	Zeta potential (mV)	Particle stability test	Cell line(s)	In situ cytotoxic assay	Incubation time (h)	Cell density (cells/ cm^2)[15]	LD_{50} (μg/mL)
Moulton et al. [41]	11.5–24.2 (i), 3.8–91.3 (ii)	(i) Epicatechin (ii) Tea extract	0.1000	NP	Intracellular AgNP ~24 h [ultra-resolution imaging], AgNP in buffer and cell media [DLS]	Keratinocyte (HeCaT)	1. MTS colorimetric assay[3] 2. LDH assay[9]	24 (1)	NR	No significant toxicity observed at tested concentrations
Powers et al. [42]	6 (i), 21 (ii), 75 (iii) [TEM]	(i) Citrate (ii) PVP	0.0010	NP	DLS, Plasma-optical emission spectroscopy	PC12	Trypan blue exclusion assay[9]	24	NR	LD_{50} levels are not reported
Trickler et al. [43]	28.3 ± 9.6, 47.5 ± 5.6, 102 ± 32.8 [TEM]	PVP	NP	−44.2, −46.0, −29.5	AgNP in suspension in cell media [DLS, LDV—laser Doppler velocimetry]	Primary rat brain microvessel endothelial (rBMEC)	XTT colorimetric assay[3]	24	4,808	No significant toxicity observed at tested concentrations
Arora et al. [44]	7.0–20.0[1] [HR-TEM]	DMEM [AgNP resuspension]	NP	NP	Intracellular AgNP [TEM], Dispersion of AgNP in cell media [DLS]	(a) Primary fibroblast from Swiss albino mice (b) Primary liver cells from Swiss albino mice	1. XTT assay[3] 2. Apoptotic/necrosis assay	24	9,615 (1), 6,172 (2)	In-situ cytotoxicity test (1). (a) 61.00 (b) 449.0

(continued)

Table 2 (continued)

Author	Size (nm) [Measurements]	Capping agent	Initial [Ag]/(M)	Zeta potential (mV)	Particle stability test	Cell line(s)	In situ cytotoxic assay	Incubation time (h)	Cell density (cells/ cm^2)[15]	LD_{50} (μg/mL)
Travan et al. [45]	33.0 ± 7.6 [TEM]	Chitlac	0.0005–0.0010	NP	~90 days [UV-Vis]	(a) Mouse fibroblast-like (NIH-3T3) (b) Human Hepatocarcinoma (HepG2) (c) Human Osteosarcoma (MG63)	1. MTT colorimetric assay[3] 2. LDH assay[9]	24, 72 (1)	4,808 (1), 350,000 (2)	No significant toxicity observed at tested concentrations
AshaRani et al. [46]	6.0–20.0 [TEM]	Starch	0.0010	NP	AgNP in cell media ~48 h [TEM]	(a) Normal human lung fibroblast (IMR-90) (b) Human glioblastoma (U251)	1. Cell-titer glow luminescent assay[14] 2. Mitochondrial—Cell titer blue assay[7] 3. Cell cycle analysis[12] 4. Apoptosis/necrosis assay	24, 48, 72	9,615 (1–2)	No significant toxicity observed at tested concentrations

In some cases LD50 values were estimated from the experimental data presented in the articles, see footnote
Cell viability assays are explained in 2.4
NR Not reported; *NP* not provided; [1]AgNPs supplied by a manufacturer; [2]AgNPs were not characterized further; [3]tetrazolium derivatives that uses the reduction properties to measure redox activity (see 2.4.1); [4]stained with calcein-AM and ethidium homodimer-1 to indicate intracellular esterase activity and membrane integrity; [5]primary osteoblast growth and differentiation measured using enzymatic alkaline phosphatase; [6]primary osteoclast growth and differentiation measured using enzymatic acid phosphatase; [7]resazurin salt that uses the reduction properties to measure metabolic mitochondrial redox properties (see 2.4.1); [8]cell lysosome marker; [9]measures membrane permeability/integrity (see 2.4.1); [10]commassie blue staining was employed to measured total protein content; [11]cell viability marker that fluoresces upon esterase activity; [12]via propidium iodide staining; [13]via Annexin V (early apoptosis)/Propidium Iodide, Hoechst, TMRE staining; [14]measure metabolic activity based on ATP concentrations; [15]Cell densities were calculated using the reported information concerning the number of cells and the plates used in each case. In some cases either cell density or plates used was not included, thus we could not calculate cell density values

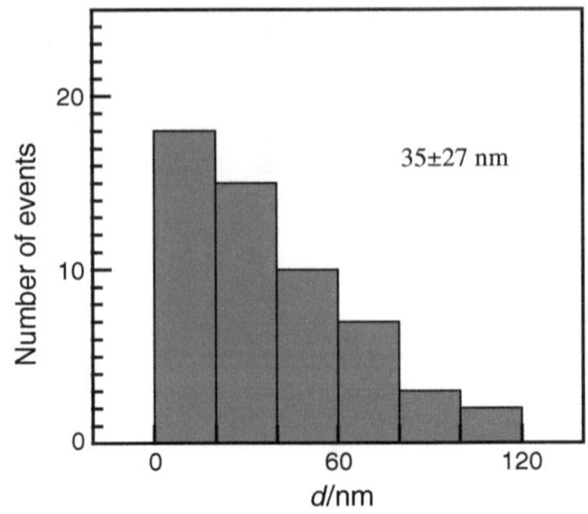

Fig. 2 Size distribution histogram for AgNP obtained from TEM images. The bin size was fixed to 20 and the numbers in the plot correspond to the mean size and standard deviation. Total number of points was 55

Note that in the calculations above we have not considered differences in nanoparticle uptake derived from differences in cell metabolism that have been reported to influence nanoparticle uptake [47]. Thus, for example, the cell uptake for fluorescently labeled carboxylated polystyrene nanoparticles can be ranked as $G_2/M > S > G_0/G_1$, with internalized particles being split with daughter cells after cell division.

2.2 Nature of the Capping/Protecting Agent

The idea of using unprotected, *a.k.a.* "naked" nanoparticles in any form of solution, for silver nanoparticles is an unrealistic goal. Silver in metallic state is readily oxidized by dissolved oxygen traces in the solution. Thus, the use of protecting or capping agents in the preparation of AgNP is critical. The role of such protecting agents, however, goes beyond merely preventing surface oxidation. In fact, the capping agent dictates the surface charge for the nanoparticle. This charge controls for the formation of complexes with biomolecules and eventually cell-nanoparticle interaction [48, 49], as a solution of nanoparticles is considered as a colloidal solution whose stability depends on the formal charge of the second hydration sphere of the nanoparticle surface [50]. The formal charge on a single nanoparticle will dictate the tendency of two particles to come in close contact to form aggregates. Thus, for example, the probability of two particles with the same charge coming close decreases at larger surface charges (i.e. zeta potential). Nanoparticle solutions with formal charges of zero are classified under the category of colloidal suspensions, while nanoparticle solutions with formal charges larger than ±20 mV are considered stable and meet the criteria for colloidal solutions.

Although the interaction between the capping agent and silver nanoparticles will vary from case to case, some lessons can be learned by examining the interaction(s) between ionic silver and biological macromolecules. The use of silver staining in gel electrophoresis, for example, is one of the best examples for the interaction between silver and biomolecules. This is an efficient staining protocol that allows biomolecule detection at the nanogram scale and is fully compatible with mass spectroscopy experiments [51, 52]. Although a complete review of the interactions between proteins and ionic silver goes beyond the scope of this chapter, in the following we will briefly review those interactions and discuss how they can relate to the interaction of silver nanoparticles and proteins.

Silver (I), Ag^+, is considered a soft cation, and has an electronic structure similar to that of copper (I), with an outer d^{10} (3d) orbital shell structure. This electronic configuration is responsible for the lack of color of Ag^+ in solution and also controls the affinity of this cation to bind biologically relevant molecules as the endogenously Cu^+ does. The prevalence of ionic silver in solution will be primarily determined by the pH and presence of polarizable atoms with which the cation forms strong complexes. Figure 3 shows the solubility diagram versus pH for different Ag species, including Ag^+. From this diagram it can be clearly seen that ionic silver is stable within pH 0 and 8. Further increment in pH produces a decrease in the solubility of the cation, probably due the formation of silver hydroxide, see Eq. 6.

$$2Ag^+ + 2OH^- \rightarrow Ag_2O\ (s) + H_2O \tag{6}$$

Ionic silver readily forms complexes with cyanide (CN^-) and chloride (Cl^-) ions. Those complexes although different in colors, produce a considerable reduction in the oxidation potential of the silver cation. Further addition of CN^-, for example, leads the formation of $Ag(CN)_2^-$, see Eqs. 7 and 8, which can be oxidized by dissolved oxygen in the solution.

$$Ag^+ + CN^- \rightarrow AgCN\ (s) \tag{7}$$

$$AgCN\ (s) + CN^- \rightarrow Ag(CN)_2^- \tag{8}$$

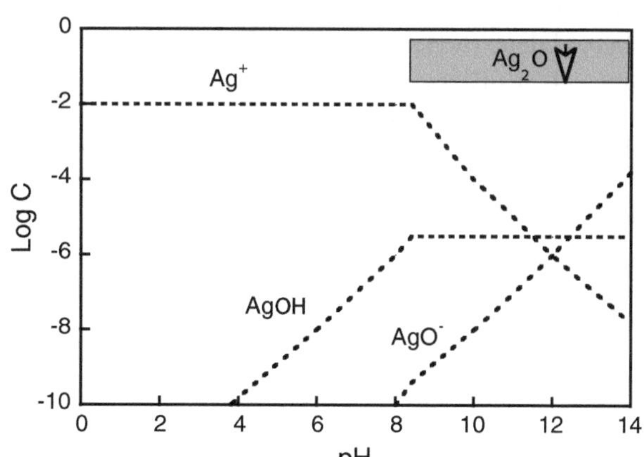

Fig. 3 Log C versus pH diagram for different species of silver in aqueous solutions. The *arrow* shown for Ag_2O indicates compound precipitation. Diagram was obtained at 10 mM silver concentration. Adapted from Burriel et al. [85]

Table 3 Complex formation constants for ionic silver and different anions and/or molecules measured at 25 °C and at the limit of ideal solutions, see footnote

	Log b$_1$	Log b$_2$	Log b$_3$	Log b$_4$
Ag^+/Br^-	4.7	7.7	8.7	9.0
Ag^+/CN^-		20.5	21.4	21.8
Ag^+/Cl^-	3.3	5.3	6.0	6.0
Ag^+/I^-	6.6	11.7	13.1	14.2
Ag^+/IO_3^-	0.6	1.9		
Ag^+/NO_2^-	2.3	2.5		
Ag^+/NH_3	3.3	7.2		
Ag^+/HS*	13.6			
$Ag^+/2HS$*		17.7		

Values listed in the table corresponds to the ionic pair equilibrium constant formation or * to the complex formation association constant defined as $\beta_n = \frac{[ML_n]}{[M][L]^n}$ where n represents the stoichiometry for the complex formation

Other examples for complex formation of Ag^+ include sodium carbonate, phosphate, and ammonium as shown in Eqs. 9–11. Although a detailed discussion about the ability of ionic silver goes beyond the scope of this chapter, Table 3 includes a collection of relevant association constants for complex formation of Ag^+ and ions naturally present in biological fluids.

$$2Ag^+ + CO_3^{-2} \rightarrow Ag_2CO_3\,(s) \tag{9}$$

$$3Ag^+ + PO_3^{-2} \rightarrow Ag_3PO_4\,(s) \tag{10}$$

$$Ag^+ + NH_3 \rightarrow Ag(NH_3)_2^+ \tag{11}$$

The intrinsic ability of ionic silver to form a complex with thiols has been used, for example, in colorimetric detection using the archetypical p-dimethylaminobenzylidene rhodanine (pRho) dye. Figure 4 shows the proposed structure for the ionic silver-pRho complex, which has a characteristic violet color.

In particular, the interactions of ionic silver and the atoms depicted in Fig. 5 plus the ability of this ion to form back-donating complexes with soft atoms

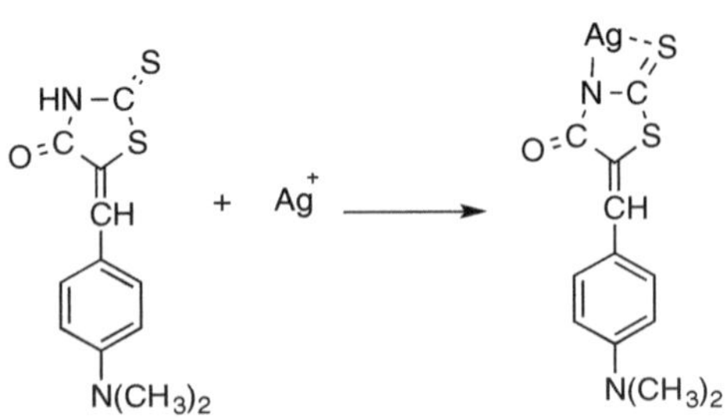

Fig. 4 Proposed chemical structure for p-dimethylaminobenzylidene rhodanine-silver complex formed in acidic solutions

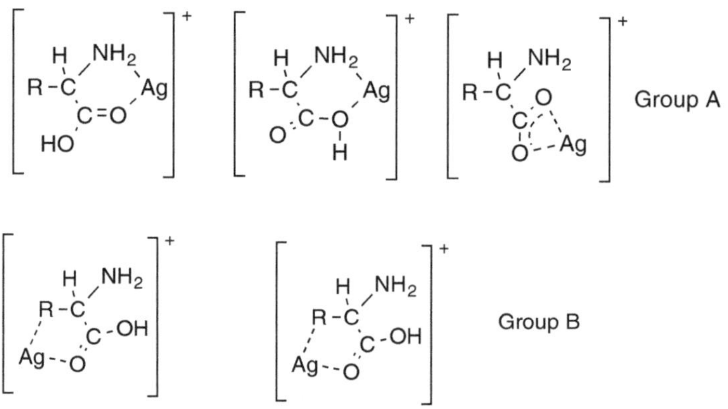

Fig. 5 Proposed chemical structures for complexes formed between amino acid residues and silver. *Group A* corresponds to bidentate complexes that do not involve participation of the side chain (R), although the participation of oxygen and nitrogen atoms in the complex formation is crucial. *Group B* shows the formation of bidentate complexes for silver with the participation of side chain residue. The possibility of forming a tridentate complex for *Group A* with the involvement of the side chain is also possible. Figure adapted from reference [53]

like oxygen play a pyramidal role in their interaction with amino acids, and by extension proteins [53]. Figure 5 contains some of the potential structures for the complexes formed between Ag^+ and amino acids/proteins.

Density functional theory (DFT) calculations and experimental measurements for the affinity of amino acids to ionic silver agree in the relevance of side chains amine groups in the complex formation [54, 55]. However, despite the natural interaction between ionic silver and thiols, studies on the thermodynamic data for the complex formation between Ag^+ and cysteine are not currently available. Regardless, AgNP had been used for the detection of cysteine in biological fluids due the high affinity that silver has for the –SH group in cysteine [56].

2.3 Nanoparticle Stability in the Chosen Cell Culture Media or Biological Fluids

All the above mentioned concerning the role of the capping agent, and the importance of having stable colloidal solutions of nanoparticles for a reliable data interpretation, becomes more crucial in complex solutions, like cell culture media or biological fluids. Table 4 contains a list of the inorganic salt composition for some of the most commonly used media in cell culture. As discussed in the previous paragraphs, since the surface composition of the nanoparticle is the immediate factor that dictates the behavior of the nanoparticle with its surrounding, changes in ionic strength, for example, will produce significant modifications in the hydrodynamic diameter of the nanoparticle. Those changes will change the nanoparticle affinity for

Table 4 Inorganic salt composition for selected cell culture media expressed in mM concentrations

Component	MEM[a] [57]	DMEM [58]	F12 [58, 59]	HBBS [60]
$CaCl_2$	1.8	1.8	0.3 ($2H_2O$)	1.3
KCl	5.4	5.4	3.0	5.4
$MgSO_4$	–	0.8	–	0.8 ($7H_2O$)
NaCl	116	109.4	130	137
$NaHCO_3$	23.8	44	14	–
NaH_2PO_4	1.1 ($2H_2O$)	0.9	–	–
$ZnSO_4$	–	–	0.003 ($7H_2O$)	–
Na_2HPO_4	–	–	1.0 ($7H_2O$)	0.33 ($2H_2O$)
$MgCl_2$	1.0 ($6H_20$)	–	0.6 ($6H_20$)	
$Fe(NO_3)_3$		0.00025 ($9H_2O$)	–	–
$CuSO_4$	–	–	0.0025 ($7H_2O$)	–
$FeSO_4$	–	–	0.003 ($7H_2O$)	–
KH_2PO_4	–	–	–	0.44

[a]In brackets are reported the hydration water molecules for the inorganic salts reported by the authors in their original papers

the cell membrane, formation of protein corona, to nanoparticle aggregation [48]. Thus, modifications on the nanoparticle stability in a given cell culture media should be assessed before cell experiments are carried out to guarantee the preservation of the material as a stable colloidal solution during the duration of the experiment.

A representative example of the importance of evaluating the stability of AgNP in high ionic strength medium is shown in Fig. 6, where it can be seen that the nature of the capping agent plays a determining role in the stability of the nanoparticle. Thus, for example in the case of citrate protected AgNP (citrate@AgNP), they were not stable in any of the tested media as revealed in the decrease of the surface plasmon band of the AgNP. However, pre-incubation with human serum albumin (HSA) of citrate@AgNP increased the stability of the nanomaterial due to a capping agent exchange. In Table 2, we have also included a column regarding the evaluation of the nanomaterial stability, where it can be seen that only a few articles had effectively carried out such experiments, despite the fact they are extremely important for the interpretation of the results concerning the toxicity of AgNP in cells.

2.4 Experimental Differences in the Cell Viability Assays

As diverse as the nature of the capping agents employed for the synthesis of AgNP are, the methodologies used for evaluating cell toxicity of AgNP also are as seen in Table 1. Although the specific list for those methodologies is quite long, they can be grouped in the following main categories:

i. *Cell membrane markers/dyes*
ii. *Redox activity sensors*
iii. *Cell proliferation assays*

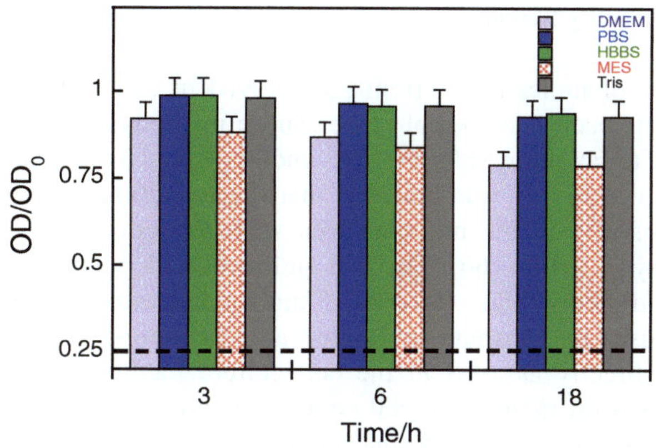

Fig. 6 Effect of 5.0 μM HSA on AgNP plasmonic absorption evaluated in different media; DMEM, PBS, HBBS, MES, and Tris-HCl buffer at different time points of incubation. OD_0 corresponds to the optical density at the SPB maximum at time zero. The *dashed line* indicates the mean OD/OD_0 value obtained for citrate@AgNP without the addition of HSA in the different media. All measurements were carried out at room temperature in quadruplicate, figure adapted from reference [15]

2.4.1 Cell Membrane Markers

Integrity of the cell membrane has been used as an archetypical marker for cell toxicity in cell biology. The membranes of healthy cells are impermeable to a range of dyes, i.e. these dyes are excluded from live cells. Dying or dead cells no longer have intact membrane, allowing these dyes to enter. The most common and widely used dye is trypan blue [61], which stains dead cells with an intense blue color. Propidium iodide is a fluorescent molecule that is also membrane impermeant to healthy cells, entering dead cells to bind nucleic acids and thereby enhance its fluorescence, which can be visualized. Similarly, ethidium bromide, a component of "live/dead" kits, is intercalated by DNA in dead cells leading to fluorescence.

Similarly, lactate dehydrogenase (LDH) leaking upon loss of the membrane integrity from the cytosol had been also widely employed as a toxicity marker [62, 63]. Note that the LDH content can also be employed as a cell viability marker in tissues since this is stable up to 48 h after cell death [64, 65]. Live/dead® fluorescent assay, on the other hand, uses the increment in red fluorescence of cell nucleus due to the DNA intercalation of ethidium bromide homodimer. Thus, the presence of cells with a red fluorescent nucleus is considered as a marker for cell toxicity, while non-stained cells are considered as viable or healthy ones, since Live/Dead® contains Calcein-AM which is hydrolyzed by intracellular esterases leading the formation of the green fluorescent dye.

2.4.2 Redox Activity Sensors

Among the list of molecules currently employed in cell viability experiments, tetrazolium salts occupy a special place. Those methods are based on the color change of the tetrazolium salts as they undergo reduction in metabolic active cells. Although at first it was believed that the reduction of those compounds occurred exclusively in the mitochondria, such reduction can also occur as a product of enzyme action [66]. The first and most widely employed tetrazolium salt is 3-(4,5-dimethylthiazo l-2-yl)-2,5-diphenyltetrazolium bromide (MTT), that was first used by Mosman to develop a cell counting system in which cell viability was directly assayed in the cell culture medium [67], with remarkable correlation with radioactive cell counting assays [68]. However, due to the reduced water solubility of the formazan compound formed upon reduction, the use of organic solvents and/or surfactants is required prior absorbance measurements are carried out. This ultimately led to the development of new tetrazolium derivatives with improved water solubility in their reduced form. MTS, XTT, WST-1 [69] and WST-8 [70]. Despite the improved sensitivity and cell compatibility displayed by those reagents, they are affected by the presence of chemicals like ascorbic acid or mercaptoethanol, something not observed for MTT [69]. Note that changes in cell metabolism and/or chemical agents, like AgNP protecting agents, present in the complex mixture can compromise the reliability of the experimental data.

The resazurin is a weakly fluorescent dye that upon reduction forms resorufin; a highly fluorescent compound. This reduction occurs inside the cell and several enzymes are believed to be involved, including cytochromes, flavin mono and di-nucleotide, flavin adenine dinucleotide, and nicotinamide adenine dehydrogenases [71]. Further, both cytosolic and microsomal enzymes have proven to be able to reduce resazurin [72]. After reduction, resorufin is excreted to the medium resulting in the pink color of the supernatant. Although changes in absorbance can be also employed for the resazurin assay, the use of the changes in fluorescence emission provides more reliable and sensitive measurements that can obtain measurements from as few as 50 cells in a 96 well plate [73].

In vitro studies using the neutral red dye (NR) were initially developed for assessment of viral cytopathogenicity and for the detection of toxic compounds at the Rockefeller University [74]. This assay is based on the ability of viable cells to incorporate and bind NR, which is mainly located in the lysosomes. Thus, the cells are lysed and the incorporated NR dye is released to the solution, giving its characteristic red color [75]. Since the dye incorporation and retention are markedly affected by the integrity of the cell membrane and pH of the lysosomes, NR assay can be used to discern between dead and damaged cells [74]. As part of the end-point assays, NR is not compatible with other toxicity assays, since the sample will be sacrificed for releasing the NR which is contained inside the cells.

2.4.3 Cell Proliferation Assays

In vitro cell proliferation is one of the most accurate tools for determining long-term effects of AgNP, like cell cycle arresting and modifications in cell proliferation. Numerous methodologies can be used to this end, including redox activity sensors, DNA quantification, protein excretion, incorporation of BrdU; a thymidine analog during DNA synthesis, and finally manual counting of cells at different time intervals. From that list, the first two assays do not allow the use of the same sample for the whole duration of the experiment. Of relatively low cost and ease of implementation is cell counting at different time points after addition of AgNP, which has proven to render relevant information regarding long-term effects of AgNP in primary cell cultures of human skin fibroblasts [9, 20]. However, this assay is not recommended for cells that either grow in suspension (HL60) or that proliferate following an island-growth pattern (HeLa). Other varieties of cell proliferation assays include, for example, the measurement of enzymatic activity like alkaline phosphatase assays [76], metabolic reporter assays like those based on ATP production like CellTiter-Glo® Luminescent Cell Viability Assay, and flow cytometry-based cell cycle analysis [77].

The above-mentioned remarks highlight the importance of consciously choosing the toxicity assay where the final intended application for AgNP must be carefully considered in the decision. The large data dispersion shown in Fig. 7 shows for the LD50 (μg/L) values for different sizes of spherical AgNP, despite the intrinsic differences between cell lines, a representative example of the inconsistences regarding AgNP toxicity found in the literature. This figure, for example, shows that for a given nanoparticle diameter i.e. 50 nm LD_{50} values range from 10 to 350 μg/mL, while for 4 nm AgNP their LD_{50} values vary from 0.11 to 1.5 μg/mL. As we have discussed in this chapter, the reasons for those differences are numerous, but certainly the lack of

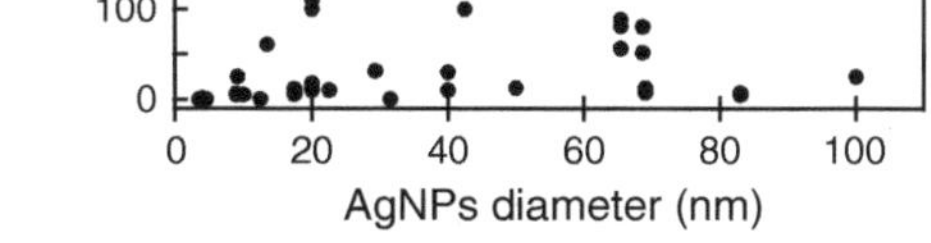

Fig. 7 In-vitro toxicity of AgNP expressed as LD_{50} values in μg/mL measured for *spherical* nanoparticles with different average diameters. Original data is presented in Table 2. Note that in many cases LD_{50} values were not directly presented, so we had to manually estimate them from the available experimental data shown in the publications

standard protocols suitable to guarantee a more direct comparison for the in vitro toxicity experiments does not make the labor any easier for scientists. In general terms, one can argue that in vitro toxicity experiments for AgNP intended for translational applications should be limited to use of primary human cultures, leaving aside cancer and animal cells, which have been widely employed due to their relative easy/low cost culture. However, extrapolation from that data, particularly for cancer cells, in terms of AgNP toxicity is not accurate and certainly had played a critical role in the existing discrepancy on AgNP toxicity.

3 Biomedical Devices and Silver Nanoparticles

A medical device is a term that defines a variety of products (excluding drugs) that can be used to aid the treatment, mitigation, diagnosis, and prevention of a disease or medical condition [78]. The U.S. Food and Drug Administration (FDA) has developed a method of classifying over thousands of medical devices used in the industry, sorting various devices under sixteen groups of various medical specialties ranging from anesthesiology to toxicology [79]. Those devices are grouped into different classes I, II, or III (Fig. 8), scored from the lowest to the highest risk to the patient. As the risk increases so does the list of criteria that a device must meet to qualify in that category [79]. Thus, for example, a class I medical device needs the lowest regulatory normative and this is intended for low to moderate risk medical procedures [79]. A class II device is subject to much higher regulatory

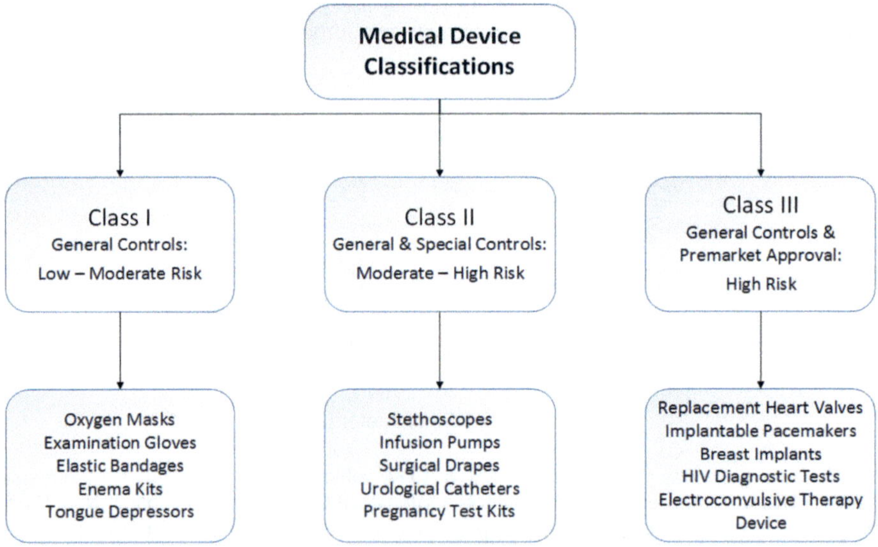

Fig. 8 Medical device classification system according to the Food and Drugs Administration (FDA)

controls and it is used for moderate to high-risk medical applications. Finally, a class III device is intended for high-risk medical applications and it has to first obtain premarket regulatory approval [79]. Devices under this classification are designed and/or intended to sustain life and/or prevent fatal injuries [79]. Although the health and sanitary regulations vary from country to country, in general they follow well-established international norms. Thus, for example Canada classifies medical devices according to the European Council Directive for devices. This system is similar to the FDA's classification used in the United States; however in Canada we have implemented a fourth class for the highest risk medical devices [80]. The therapeutic products directorate (TPD) is the governing body in Canada in charge of monitoring and evaluating the safety and effectiveness of medical devices through assurance of pre-market review, post-approval surveillance, and quality systems [78].

3.1 Silver Nanoparticles in Biomedical Devices

In the following we will briefly describe some examples of medical devices from the different classes that contain AgNP.

(i) *Dental Material* (*Class I*) Studies have shown that AgNP infused dental filling cements have remarkable bactericidal effects [81]. Adding AgNP to orthodontic adhesives has also demonstrated notable antimicrobial properties, which in turn prevents enamel demineralization without jeopardizing the physical attributes of the adhesive in the process [82].

(ii) *Wound Dressings* (*Class I–Class II*) Recent studies have shown that wound dressings containing AgNP can be used to inhibit infection and accelerate the healing process as a result of the reduced bacteria colonization [83, 84]. Acticoat™ was the first commercially available product utilized for wound dressing containing AgNP [85]. This product was clinically tested for skin wounds including burns, Steven-Johnson syndrome, chronic ulcers, and pemphigus [84, 86, 87].

(iii) *Contact Lens* (*Class II*) With the increasing demand for contact lenses, the antibacterial effects of silver for the development of long-term-use contact lenses have experienced a revived interest. Studies have shown that the incorporation of ionic silver in contact lens casings confers antibacterial properties to the device [88]. In fact, contact lens cases containing ionic silver have been approved by the FDA and are currently commercially available (CIBA Vision Corp., GA, USA; Sauflon Pharmaceuticals, Twickenham, UK). In vitro studies have also shown that AgNP incorporated within contact lenses have reduced the bacterial colonization of the implant [89]. However, in the case of AgNP, further clinical testing is required prior commercially viable product hits the market.

(iv) *Medical Catheters* (*Class II*) The antibacterial effect of AgNP on polyurethane, one of the most common polymers used in manufacturing medical catheters, has been widely studied in the past decade as an application that inhibits biofilm formation and prevents catheter indwelling on central venous and neurosurgical catheters [90]. In vitro, in vivo and clinical studies have shown the effective inhibition of biofilm development along with excellent biocompatibility [91–93]. Silverline® [94] and ON-Q Silver Soaker™ are some of these commercially available AgNP containing catheters in the market that currently exemplify a practical use of the innate antibacterial properties AgNP.

(v) *Bone Cement* (*Class II*) AgNP integrated poly (methyl methacrylate) (PMMA) based bone cement used for the secure attachment of artificial joint replacements have demonstrated notable prophylactic effects along with exceptional biocompatibility [95]. AgNP contained within an ultra-high-molecular-weight-polyethylene, a product used as inserts for joint replacement, have been shown to drastically overcome the wear and tear effect of this polymer [96].

(vi) *Surgical Meshes* (*Class II–Class III*) The rationale behind using surgical meshes is to create a connecting conduit for substantial open wounds during surgical procedures, which assists in decreasing the amount of postoperative nosocomial infections [97]. One study demonstrated that AgNP integrated polypropylene surgical meshes have a sustained bactericidal effect [98].

(vii) *Cardiovascular Implants* (*Class III*) A silicone prosthetic heart valve containing elemental Ag was designed to both prevent bacterial infections and reduce the effects to the inflammatory process [99]. Silzone® was the first commercial Ag coated heart valve available which exhibited promising biocompatibility [100]. Unfortunately, after several years on the market the product was recalled because it had caused paravalvular leakage in clients through the inhibition of normal fibroblast functioning, triggering hypersensitivity. Recent studies have looked at using AgNP and carbon coated thin films as a potential alternative in heart valves as they have demonstrated anti-thrombogenic properties [101].

4 Silver Nanoparticles in Advanced Therapy Medicinal Products

Advanced Therapy Medicinal Products (ATMP) are therapies that are generally destined for regenerative medicine. There are several therapies that fit under the ATMP definition. These include gene therapy products, somatic cell therapy products, and tissue engineered products. AgNP have been used as part of complexes for gene delivery, for example, photoactivated microRNA delivery to adipose-derived mesenchymal stem cells, to promote their differentiation into osteogenic or bone lineages [102].

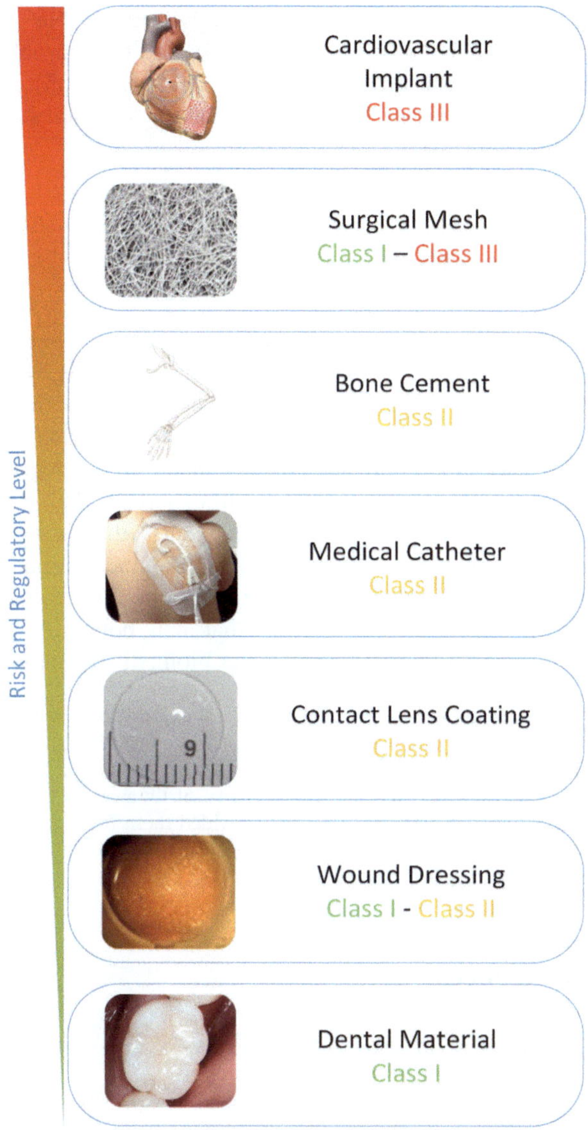

Fig. 9 Various examples of AgNP based medical devices and their designated class type

In tissue-engineered products, silver has generally been used as an anti-bacterial additive. However, there are a few reports where AgNP served as active components that modulated cell behavior and differentiation. For example, AgNP were used to stimulate the differentiation of urine-derived stem cells into osteogenic cells, by up-regulating the expression of osteogenesis-related genes [103]. Similar effects were reported for mesenchymal stem cells that were included to differentiate into bone cells by AgNP [104]. In other studies, AgNP were found to modulate the intracellular signaling pathways to enable the neuronal differentiation. Although this early work was performed on neuroblastoma cells, further research could lead to utility with neuronal stem cells [105]. However, caution is needed as AgNP cytotoxicity has been reported for neuronal stem cells [106] as well as other progenitors [107].

5 Future Perspectives

With substantial technological advancements being made for various medical devices, the use of nanomaterials including AgNP has gained a renewed vitality. However, prior to any realistic use of AgNP in ATMP or implantable medical devices, issues concerning their long-term safety and biocompatibility must be addressed in a timely and proper manner. In this regard, advancements in nanotechnology, particularly surface modification for AgNP achieved in the past 5 years, aim to overcome toxic side effects derived from AgNP oxidation, aggregation or ionic silver release, and aim to further reduce immune response and implant rejections. Thus, a whole new horizon of possibilities remains to be explored in uses and applications, possibilities that will hopefully lead to the development and implementation of safer and better biomedical devices containing AgNP.

Acknowledgments We would like to express our deeply gratitude to all the researches who have been cited in this chapter. EIA thanks the University of Ottawa Heart Institute for the financial and scientific support (UOHI grant #1255). The authors would like to thank Professor Juan C. Scaiano at University of Ottawa for his permanent and generous Scientific and Professional guidance he had provided them for many years in the field of nanomaterials. We would like to thank Dr. George Wehbi, Elgin Dental Care and Mr. Bastian Leon for kindly supplying images for Fig. 9.

References

1. Hill, W.R., Pillsbury, D.M.: Argyria-the pharmacology of silver. J. Am. Pharm. Assoc. **29**(5), 239–240 (1940)
2. Alexander, J.W.: History of the medical use of silver. Surg. Infect. **10**(3), 289–292 (2009)
3. Varner, K.E., et al.: State-of-the-science review: everything nanosilver and more. U.S. Environmental Protection Agency, Washington, DC (2010)
4. Credé, C.S.F.: Die Verhutung der Augenentzundung der Neugeborenen (ophthalmoblennorrhoea neonatorum), der haufigsten und wichtigsten Ursache de Blindheit. Verlag von August Hirschwald, Berlin (1884)
5. Klasen, H.J.: A historical review of the use of silver in the treatment of burns. II. Renewed interest for silver. Burns **26**(2), 131–138 (2000)
6. Vecki, V.G., Ottinger, M.R.: Treatment of syphilis with silver-salvarsan. Calif. State J. Med. **19**(11), 438 (1921)
7. Lansdown, A.B.G.: Chapter 1 silver in health and disease. In: Lansdown, A.B.G. (ed.) Silver in Healthcare: Its Antimicrobial Efficacy and Safety in Use. The Royal Society of Chemistry, Cambridge, pp. 1–8 (2010)

8. Aziz, Z., Abu, S.F., Chong, N.J.: A systematic review of silver-containing dressings and topical silver agents (used with dressings) for burn wounds. Burns **38**(3), 307–318 (2012)
9. Vignoni, M., et al.: LL37 peptide@silver nanoparticles: combining the best of the two worlds for skin infection control. Nanoscale **6**(11), 5725–5728 (2014)
10. Gamelli, R.L., Paxton, T.P., O'Reilly, M.: Bone marrow toxicity by silver sulfadiazine. Surg. Gynecol. Obstet. **177**(2), 115–120 (1993)
11. Agency, U.S.E.P. Silver (CASRN 7440-22-4). http://www.epa.gov/iris/subst/0099.htm (1987). Accessed 10 Oct 2014
12. Pacioni, N.L., et al.: Surface plasmons control the dynamics of excited triplet states in the presence of gold nanoparticles. J. Am. Chem. Soc. **132**(18), 6298–6299 (2010)
13. Chithrani, B.D., Ghazani, A.A., Chan, W.C.W.: Determining the size and shape dependence of gold nanoparticle uptake into mammalian cells. Nano Lett. **6**(4), 662–668 (2006)
14. He, C., et al.: Effects of particle size and surface charge on cellular uptake and biodistribution of polymeric nanoparticles. Biomaterials **31**(13), 3657–3666 (2010)
15. Alarcon, E.I., et al.: Human serum albumin as protecting agent of silver nanoparticles: role of the protein conformation and amine groups in the nanoparticle stabilization. J. Nanopart. Res. **15**(1), 1374–1377 (2013)
16. Albers, C.E., et al.: In vitro cytotoxicity of silver nanoparticles on osteoblasts and osteoclasts at antibacterial concentrations. Nanotoxicology **7**(1), 30–36 (2013)
17. Ashraf, S., et al.: Protein-mediated synthesis, pH-induced reversible agglomeration, toxicity and cellular interaction of silver nanoparticles. Colloids Surf. B **102**, 511–518 (2013)
18. Grosse, S., Evje, L., Syversen, T.: Silver nanoparticle-induced cytotoxicity in rat brain endothelial cell culture. Toxicol. In Vitro **27**(1), 305–313 (2013)
19. Kermanizadeh, A., et al.: In vitro assessment of engineered nanomaterials using a hepatocyte cell line: cytotoxicity, pro-inflammatory cytokines and functional markers. Nanotoxicology **7**(3), 301–313 (2013)
20. Simpson, M.J., et al.: Impact of dye-protein interaction and silver nanoparticles on rose Bengal photophysical behavior and protein photocrosslinking. Photochem. Photobiol. **89**(6), 1433–1441 (2013)
21. Suman, T.Y., et al.: Biosynthesis, characterization and cytotoxic effect of plant mediated silver nanoparticles using Morinda citrifolia root extract. Colloids Surf. B **106**, 74–78 (2013)
22. Alarcon, E.I., et al.: The biocompatibility and antibacterial properties of collagen-stabilized, photochemically prepared silver nanoparticles. Biomaterials **33**(19), 4947–4956 (2012)
23. Asare, N., et al.: Cytotoxic and genotoxic effects of silver nanoparticles in testicular cells. Toxicology **291**(1–3), 65–72 (2012)
24. Haase, A., et al.: Effects of silver nanoparticles on primary mixed neural cell cultures: uptake, oxidative stress and acute calcium responses. Toxicol. Sci. **126**(2), 457–468 (2012)
25. Hayashi, Y., et al.: Earthworms and humans in vitro: characterizing evolutionarily conserved stress and immune responses to silver nanoparticles. Environ. Sci. Technol. **46**(7), 4166–4173 (2012)
26. Justin Packia Jacob, S., Finub, J.S., Narayanan, A.: Synthesis of silver nanoparticles using Piper longum leaf extracts and its cytotoxic activity against Hep-2 cell line. Colloids Surf. B **91**(0), 212–214 (2012)
27. Lu, R., et al.: Egg white-mediated green synthesis of silver nanoparticles with excellent biocompatibility and enhanced radiation effects on cancer cells. Int. J. Nanomed. **7**, 2101 (2012)
28. Mukherjee, S.G., et al.: Comparative in vitro cytotoxicity study of silver nanoparticle on two mammalian cell lines. Toxicol. In Vitro **26**(2), 238–251 (2012)
29. Suresh, A.K., et al.: Cytotoxicity induced by engineered silver nanocrystallites is dependent on surface coatings and cell types. Langmuir **28**(5), 2727–2735 (2012)
30. Bouwmeester, H., et al.: Characterization of translocation of silver nanoparticles and effects on whole-genome gene expression using an in vitro intestinal epithelium coculture model. ACS Nano **5**(5), 4091–4103 (2011)

31. Comfort, K.K., et al.: Interference of silver, gold, and iron oxide nanoparticles on epidermal growth factor signal transduction in epithelial cells. ACS Nano **5**(12), 10000–10008 (2011)
32. Foldbjerg, R., Dang, D.A., Autrup, H.: Cytotoxicity and genotoxicity of silver nanoparticles in the human lung cancer cell line, A549. Arch. Toxicol. **85**(7), 743–750 (2011)
33. Haase, A., et al.: Toxicity of silver nanoparticles in human macrophages: uptake, intracellular distribution and cellular responses. J. Phys. Conf. Ser. **304**(1), 012030 (2011)
34. Hackenberg, S., et al.: Silver nanoparticles: evaluation of DNA damage, toxicity and functional impairment in human mesenchymal stem cells. Toxicol. Lett. **201**(1), 27–33 (2011)
35. Park, M.V.D.Z., et al.: The effect of particle size on the cytotoxicity, inflammation, developmental toxicity and genotoxicity of silver nanoparticles. Biomaterials **32**(36), 9810–9817 (2011)
36. Stevanović, M., et al.: Effect of poly-α, γ, L-glutamic acid as a capping agent on morphology and oxidative stress-dependent toxicity of silver nanoparticles. Int. J. Nanomed. **6**, 2837 (2011)
37. Stoehr, L.C., et al.: Shape matters: effects of silver nanospheres and wires on human alveolar epithelial cells. Part. Fibre Toxicol. **8**(36), 3–15 (2011)
38. Braydich-Stolle, L.K., et al.: Silver nanoparticles disrupt GDNF/Fyn kinase signaling in spermatogonial stem cells. Toxicol. Sci. **116**(2), 577–589 (2010)
39. Eom, H.-J., Choi, J.: p38 MAPK activation, DNA damage, cell cycle arrest and apoptosis as mechanisms of toxicity of silver nanoparticles in Jurkat T cells. Environ. Sci. Technol. **44**(21), 8337–8342 (2010)
40. Martinez-Gutierrez, F., et al.: Synthesis, characterization, and evaluation of antimicrobial and cytotoxic effect of silver and titanium nanoparticles. Nanomed. Nanotechnol. Biol. Med. **6**(5), 681–688 (2010)
41. Moulton, M.C., et al.: Synthesis, characterization and biocompatibility of green synthesized silver nanoparticles using tea polyphenols. Nanoscale **2**(5), 763–770 (2010)
42. Powers, C.M., et al.: Silver nanoparticles compromise neurodevelopment in PC 12 cells: critical contributions of silver ion, particle size, coating, and composition. Environ. Health Perspect. **119**(1), 37–44 (2010)
43. Trickler, W.J., et al.: Silver nanoparticle induced blood-brain barrier inflammation and increased permeability in primary rat brain microvessel endothelial cells. Toxicol. Sci. **118**, 160–170 (2010)
44. Arora, S., et al.: Interactions of silver nanoparticles with primary mouse fibroblasts and liver cells. Toxicol. Appl. Pharmacol. **236**(3), 310–318 (2009)
45. Travan, A., et al.: Non-cytotoxic silver nanoparticle-polysaccharide nanocomposites with antimicrobial activity. Biomacromolecules **10**(6), 1429–1435 (2009)
46. AshaRani, P., et al.: Cytotoxicity and genotoxicity of silver nanoparticles in human cells. ACS Nano **3**(2), 279–290 (2008)
47. Kim, J.A., et al.: Role of cell cycle on the cellular uptake and dilution of nanoparticles in a cell population. Nat. Nano **7**(1), 62–68 (2012)
48. Gebauer, J.S., et al.: Impact of the nanoparticle-protein corona on colloidal stability and protein structure. Langmuir **28**(25), 9673–9679 (2012)
49. Lynch, I., Dawson, K.A.: Protein-nanoparticle interactions. Nano Today **3**(1), 40–47 (2008)
50. Delgado, A.V., et al.: Measurement and interpretation of electrokinetic phenomena (IUPAC technical report). Pure Appl. Chem. **77**(10), 1753–1805 (2005)
51. Eschenbruch, M., Burk, R.R.: Experimentally improved reliability of ultrasensitive silver staining of protein in polyacrylamide gels. Anal. Biochem. **125**(1), 96–99 (1982)
52. Gharahdaghi, F., et al.: Mass spectrometric identification of proteins from silver-stained polyacrylamide gel: a method for the removal of silver ions to enhance sensitivity. Electrophoresis **20**(3), 601–605 (1999)
53. Eckhardt, S., et al.: Nanobio silver: its interactions with peptides and bacteria, and its uses in medicine. Chem. Rev. (Washington, DC, U.S.) **113**(7), 4708–4754 (2013)

54. Shoeib, T., Siu, K.W.M., Hopkinson, A.C.: Silver ion binding energies of amino acids: use of theory to assess the validity of experimental silver ion basicities obtained from the kinetic method. J. Phys. Chem. A **106**(25), 6121–6128 (2002)
55. Jover, J., Bosque, R., Sales, J.: A comparison of the binding affinity of the common amino acids with different metal cations. Dalton Trans. **45**, 6441–6453 (2008)
56. Leesutthiphonchai, W., et al.: Selective determination of homocysteine levels in human plasma using a silver nanoparticle-based colorimetric assay. Talanta **85**(2), 870–876 (2011)
57. Hay, R.: Preservation of cell-culture stocks in liquid nitrogen. TCA Man./Tissue Cult. Assoc. **4**(2), 787–790 (1978)
58. Morton, H.: A survey of commercially available tissue culture media. In Vitro **6**(2), 89–108 (1970)
59. Ham, R.G.: Clonal growth of mammalian cells in a chemically defined, synthetic medium. Proc. Natl. Acad. Sci. **53**(2), 288–293 (1965)
60. Hanks, J.H., Wallace, R.E.: Relation of oxygen and temperature in the preservation of tissues by refrigeration. Exper. Biol. Med. **71**(2), 196–200 (1949)
61. Doyle, A., Griffiths, J.B., Newell, D.G.: Cell and tissue culture: laboratory procedures. In: Doyle, A., Griffiths, J.B., Newell, D.G. (ed.) John Wiley and Sons, Inc., Chichester (1995)
62. Decker, T., Lohmann-Matthes, M.L.: A quick and simple method for the quantitation of lactate dehydrogenase release in measurements of cellular cytotoxicity and tumor necrosis factor (TNF) activity. J. Immunol. Methods **115**(1), 61–69 (1988)
63. Korzeniewski, C., Callewaert, D.M.: An enzyme-release assay for natural cytotoxicity. J. Immunol. Methods **64**(3), 313–320 (1983)
64. Wong, S.Y., et al.: The determination of bone viability: a histochemical method for identification of lactate dehydrogenase activity in osteocytes in fresh calcified and decalcified sections of human bone. Pathology **14**(4), 439–442 (1982)
65. Wong, S.Y., et al.: The effect of age on bone composition and viability in the femoral head. J. Bone Joint Surg. Am. **67**(2), 274–283 (1985)
66. Bernas, T., Dobrucki, J.: Mitochondrial and nonmitochondrial reduction of MTT: interaction of MTT with TMRE, JC-1, and NAO mitochondrial fluorescent probes. Cytometry **47**(4), 236–242 (2002)
67. Mosmann, T.: Rapid colorimetric assay for cellular growth and survival: application to proliferation and cytotoxicity assays. J. Immunol. Methods **65**(1–2), 55–63 (1983)
68. Tada, H., et al.: An improved colorimetric assay for interleukin 2. J. Immunol. Methods **93**(2), 157–165 (1986)
69. Berridge, M., et al.: The biochemical and cellular basis of cell proliferation assays that use tetrazolium salts. Biochemica **4**(1), 14–19 (1996)
70. Tominaga, H., et al.: A water-soluble tetrazolium salt useful for colorimetric cell viability assay. Anal. Commun. **36**(2), 47–50 (1999)
71. O'Brien, J., et al.: Investigation of the Alamar Blue (resazurin) fluorescent dye for the assessment of mammalian cell cytotoxicity. Eur. J. Biochem. **267**(17), 5421–5426 (2000)
72. Gonzalez, R.J., Tarloff, J.B.: Evaluation of hepatic subcellular fractions for Alamar blue and MTT reductase activity. Toxicol. In Vitro **15**(3), 257–259 (2001)
73. Czekanska, E.M.: Assessment of cell proliferation with resazurin-based fluorescent dye. In: Stoddart, M.J. (ed.) Mammalian Cell Viability. Academic Press, New York (2011)
74. Borenfreund, E., Puerner, J.: A simple quantitative procedure using monolayer cultures for cytotoxicity assays (HTD/NR-90). J. Tissue Cult. Methods **9**(1), 7–9 (1985)
75. Repetto, G., del Peso, A., Zurita, J.L.: Neutral red uptake assay for the estimation of cell viability/cytotoxicity. Nat. Protoc. **3**(7), 1125–1131 (2008)
76. Bessey, O.A., Lowry, O.H., Brock, M.J.: A method for the rapid determination of alkaline phosphates with five cubic millimeters of serum. J. Biol. Chem. **164**, 321–329 (1946)
77. Pozarowski, P., Darzynkiewicz, Z.: Analysis of cell cycle by flow cytometry. In: Schönthal, A. (ed.) Checkpoint Controls and Cancer, pp. 301–311. Humana Press, Totowa (2004)

78. Canada, H. Safe Medical Devices in Canada [Health Canada Fact Sheet]. http://www.hc-sc.gc.ca/dhp-mps/md-im/activit/fs-fi/meddevfs_matmedfd-eng.php (2014). Accessed 13 Oct 2014
79. FDA. Classify your medical device. http://www.fda.gov/%20MedicalDevices/DeviceRegulationandGuidance/Overview/ClassifyYourDevice/default.htm (2014). Accessed 13 Oct 2014
80. Justice-Canada. M.o. medical devices regulations.http://laws-lois.justice.gc.ca/eng/regulations/sor-98-282/page-2.html (2014). Accessed 13 Oct 2014
81. Magalhães, A.P.R., et al.: Nanosilver application in dental cements. ISRN Nanotech. **2012**, 365438 (2012)
82. Ahn, S.-J., et al.: Experimental antimicrobial orthodontic adhesives using nanofillers and silver nanoparticles. Dent. Mater. **25**(2), 206–213 (2009)
83. Chen, J., et al.: Effect of silver nanoparticle dressing on second degree burn wound. Zhonghua wai ke za zhi [Chin. J. Surg.] **44**(1), 50–52 (2006)
84. Huang, Y., et al.: A randomized comparative trial between Acticoat and SD-Ag in the treatment of residual burn wounds, including safety analysis. Burns **33**(2), 161–166 (2007)
85. Burriel, R.E., McIntosh, C.L., Morris, L.R.: Process of activating anti-microbial materials. Google Patents (1995)
86. Yang, J.-Y., et al.: A clinical experience of treating exfoliative wounds using nanocrystalline silver-containing dressings (Acticoat®). Burns **33**(6), 793–797 (2007)
87. Sibbald, R.G., et al.: Bacteriology, inflammation, and healing: a study of nanocrystalline silver dressings in chronic venous leg ulcers. Adv. Skin Wound Care **20**(10), 549–558 (2007)
88. Amos, C.F., George, M.D.: Clinical and laboratory testing of a silver-impregnated lens case. Contact Lens Anterior Eye **29**(5), 247–255 (2006)
89. Willcox, M.D., et al.: Ability of silver-impregnated contact lenses to control microbial growth and colonisation. J. Optom. **3**(3), 143–148 (2010)
90. Samuel, U., Guggenbichler, J.: Prevention of catheter-related infections: the potential of a new nano-silver impregnated catheter. Int. J. Antimicrob. Agents **23**, 75–78 (2004)
91. Galiano, K., et al.: Silver segregation and bacterial growth of intraventricular catheters impregnated with silver nanoparticles in cerebrospinal fluid drainages. Neurol. Res. **30**(3), 285–287 (2008)
92. Khare, M.D., et al.: Reduction of catheter-related colonisation by the use of a silver zeolite-impregnated central vascular catheter in adult critical care. J. Infect. **54**(2), 146–150 (2007)
93. Lackner, P., et al.: Efficacy of silver nanoparticles-impregnated external ventricular drain catheters in patients with acute occlusive hydrocephalus. Neurocrit. Care **8**(3), 360–365 (2008)
94. Spiegelberg, R.N.: Technology for brains|EVD-catheters. http://www.spiegelberg.de/products/drainage/evd_catheters.html (2014). Accessed 13 Oct 2014
95. Alt, V., et al.: An in vitro assessment of the antibacterial properties and cytotoxicity of nanoparticulate silver bone cement. Biomaterials **25**(18), 4383–4391 (2004)
96. Morley, K., et al.: Synthesis and characterisation of advanced UHMWPE/silver nanocomposites for biomedical applications. Eur. Polymer J. **43**(2), 307–314 (2007)
97. Darouiche, R.O.: Treatment of infections associated with surgical implants. New Engl. J. Med. **350**(14), 1422–1429 (2004)
98. Cohen, M.S., et al.: In vitro analysis of a nanocrystalline silver-coated surgical mesh. Surg. Infect. **8**(3), 397–404 (2007)
99. Grunkemeier, G.L., Jin, R., Starr, A.: Prosthetic heart valves: objective performance criteria versus randomized clinical trial. Ann. Thor. Surg. **82**(3), 776–780 (2006)
100. Jamieson, W., et al.: Seven-year results with the St Jude Medical Silzone mechanical prosthesis. J. Thorac. Cardiovasc. Surg. **137**(5), 1109–1115. e2 (2009)
101. Andara, M., et al.: Hemocompatibility of diamondlike carbon–metal composite thin films. Diam. Relat. Mater. **15**(11), 1941–1948 (2006)
102. Qureshi, A.T., et al.: miR-148b-nanoparticle conjugates for light mediated osteogenesis of human adipose stromal/stem cells. Biomaterials **34**(31), 7799–7810 (2013)

103. Qin, H., et al.: Silver nanoparticles promote osteogenic differentiation of human urine-derived stem cells at noncytotoxic concentrations. Int. J. Nanomed. **9**, 2469–2478 (2014)
104. Liu, X., et al.: Influence of silver nanoparticles on osteogenic differentiation of human mesenchymal stem cells. J. Biomed. Nanotechnol. **10**(7), 1277–1285 (2014)
105. Dayem, A.A., et al.: Biologically synthesized silver nanoparticles induce neuronal differentiation of SH-SY5Y cells via modulation of reactive oxygen species, phosphatases, and kinase signaling pathwayss. Biotechnol. J. **9**(7):934–943 (2014)
106. Soderstjerna, E., et al.: Gold- and silver nanoparticles affect the growth characteristics of human embryonic neural precursor cells. PLoS ONE **8**(3), e58211 (2013)
107. Wang, Z., et al.: Silver nanoparticles induced RNA polymerase-silver binding and RNA transcription inhibition in erythroid progenitor cells. ACS Nano **7**(5), 4171–4186 (2013)

Anti-microbiological and Anti-infective Activities of Silver

May Griffith, Klas I. Udekwu, Spyridon Gkotzis, Thien-Fah Mah and Emilio I. Alarcon

Abstract Silver nanoparticles are the latest version of silver preparations that have been revived and tested as anti-microbials, particularly as an alternative to antibiotics since the emergence of drug resistant bacteria. Silver nanoparticles share commonalities with other silver preparations and other metallic nanoparticles, but also several significant differences in their interactions with microbes. Their mechanism of action is not completely understood but their potential utility has led to the high level of research activity to determine the safety and efficacy of these nanoparticles for clinical applications.

Keywords Silver nanoparticles · Anti-infective properties

M. Griffith
Integrative Regenerative Medicine Centre,
Department of Clinical and Experimental Medicine,
Linköping University, 58185 Linköping, Sweden

M. Griffith · K.I. Udekwu · S. Gkotzis
Swedish Medical Nanoscience Centre, Department of Neuroscience,
Karolinksa Institutet, Stockholm, Sweden

T.-F. Mah
Department of Biochemistry, Microbiology and Immunology,
Faculty of Medicine, University of Ottawa, Ottawa, Canada

E.I. Alarcon (✉)
Bio-Nanomaterials Chemistry and Engineering Laboratory,
Division of Cardiac Surgery, University of Ottawa Heart Institute,
40 Ruskin Street Rm H5229, Ottawa K1Y 4W7, Canada
e-mail: ealarcon@ottawaheart.ca

E.I. Alarcon
Centre for Catalysis Research and Innovation,
University of Ottawa, Ottawa, Canada

E.I. Alarcon et al. (eds.), *Silver Nanoparticle Applications*, Engineering Materials,
DOI 10.1007/978-3-319-11262-6_6

1 Introduction and the Silver Bullet

The earliest uses of silver as an anti-microbial appears to have been from the time of the ancient Greeks and Romans. In a very comprehensive, historical review of the uses of silver by Alexander [1].

Hippocrates was credited for using silver preparations for the treatment of ulcers and to promote wound healing. Other early accounts for the uses of silver include its use internally as well as for treatment of wounds by Paracelsus in 1520, and as a treatment for brain infection in 1614 by Angelo Sala. In the 1800s, silver sutures were used in surgery were documented to allow wound healing

According to Alexander, the use of silver in medicine began to flourish, in the late 1880s, silver nitrate eye drops were applied to newborn babies to prevent gonorrheal ophthalmia, a practice that has continued well into the mid- to late 1900s. Credewas credited with being the first physician to use colloidal silver for wound antisepsis in 1890s, while Crusius used silver nitrate for the treatment of burn injuries. In the eye, Roe was documented to have used colloidal silver to successfully treat infected corneal ulcers, keratitis, blepharitis, and dacrocystitis [2].

The first person to actually connect the use of silver to its action as an anti-microbial agent is believed to be Vonnaegele [1]. He attributed the effectiveness of silver as an anti-bacterial agent to the silver ion, opening up the use of silver as an effective antimicrobial agent until the discovery of antibiotics.

While the golden era of antibiotics flourished, the use of silver as anti-microbial or anti-infective agents diminished. In the past decade, however, almost every known antibiotic has given rise to resistant strains of bacteria, and multi-drug resistant (MDR) bacteria [3]. In the quest for alternative to antibiotics, the use of the silver has been rediscovered and is coming back into mainstream medicine as an anti-microbial and anti-infective agent [4].

This chapter aims to discuss the role of silver nanoparticles as anti-infectives, focusing mainly on the interaction of silver nanoparticles with bacteria and viruses.

2 Silver and its Derivatives as Anti-microbials

2.1 Metallic Silver

Silver is a noble metal, i.e. chemically, it is inert. However, its interaction with moisture on the skin surface and with wound fluids leads to the release of silver ions, which have anti-microbial effects [5]. In biomedical applications, metallic silver has been used or tested in dental alloys (also as silver amalgam) and prosthetic implants such as "mega-endoprostheses" that are grafted after the removal of bone tumors [6], as well as silver–treated fabric for heart valves [7] and silver sutures. In bone prosthesis, incorporation of silver appears to have resulted in a reduction of the infection rate with no adverse side effects. However, the heart valves failed, and the silver sutures, which were used for artificial heart valve retention resulted in local inflammation that was attributed to released silver [8].

2.2 Silver Nitrate and Ions

Silver salts such as silver nitrate ($AgNO_3$) are one of the best known source of ionic silver. In the 18th century, ionic silver was used to treat ulcers, and in the 1960s, $AgNO_3$ was used as a topical solution (0.5 % $AgNO_3$) for management of wounds from burns (reviewed by Chopra, [9]). Silver in form of $AgNO_3$ has also been combined with antibiotics, such as sulfonamide in 1968, to produce silver sulfadiazine (SSD) cream that resulted in a broad spectrum silver-based antibacterial that continued to be prescribed mostly for the management of burns [10, 11]. The synergistic effects of AgNPs and antibiotics will be further discussed below.

The best known targets of silver ions are the thiol moieties present in the respiratory enzymes of bacterial cells [11]. Silver readily binds to the bacterial cell wall and cell membrane and evidently inhibits the respiration chain [12]. Moreover, it has been shown that in *E. coli* silver inhibits the uptake of inorganic phosphate and causes efflux of phosphate, proline, glutamine, succinate and mannitol from cells [12, 13].

2.3 Silver Nanoparticles

In general, nanoparticles can bind not only membrane bound targets but also penetrate the bacterial cell. In particular, AgNP complex strongly to electron donor groups containing N, O, S and exhibit an elevated binding constant for thiol containing proteins. AgNP preferentially attack the respiratory chain, hinder cell division and ultimately lead to cell death. Sondi and Salopek-Sondi have suggested the degradation of the membrane structure of *E. coli* during treatment with spherical 12 nm AgNP [14].

It has been shown that in order to function effectively as bactericidal agents, AgNP have to fall with a narrow size range. Silver particles made from silver atoms are exceedingly small and within the ideal range. However, these naked particles are generally unstable and are easily oxidized, see Chap. "Biomedical Uses of Silver Nanoparticles: From Roman Wine Cups to Biomedical Devices". Thus, the nature of the capping agent plays a pivotal role in the antimicrobial performance of AgNP. In Alarcon et al. [15] for example, AgNP were stabilized by coating with type I collagen or α-poly-L-lysine coated AgNP were tested against Gram positive bacteria such as *Bacillus megaterium* (Strain Bm11), a non-sporulating bacterium, and *Staphylococcus epidermis* (Strain Se4), a sporulating bacterium, as well as a Gram-negative, uropathogenic strain of *Escherichia coli* (CFT073). Both coated AgNP were found to be as effective as control Ag nitrate ($AgNO_3$) in inhibiting all three bacteria. However, it was reported that the poly-L-Lysine coated nanoparticles suffer from spontaneous aggregation at higher ionic strength or lower temperature, while the collagen-coated AgNP were very stable and also cell compatible

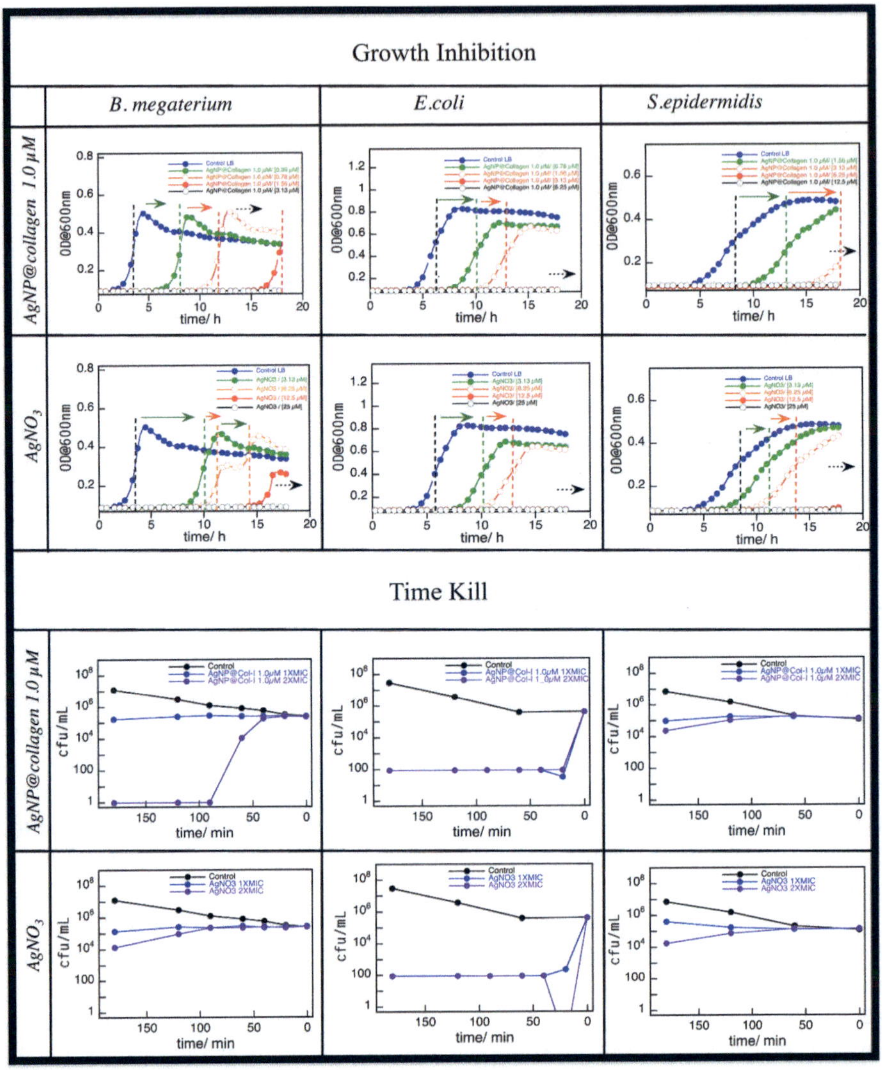

Fig. 1 *Top* Bacteria growth inhibition profile for *B. megaterium left column*, *E. coli* (*middle column*), and *S. epidermidis* (*right column*) up to 18 h in the presence of AgNP@collagen 1.0 μM and $AgNO_3$. *Dashed colored lines* indicate the time where a 50 % of the maximal OD at the plateau level. Additional black dashed horizontal arrows indicate longer lag times for all the other concentrations at >18 h. *Bottom* Bacteria time kill profiles for *B. megaterium* (*left column*), *E. coli* (*middle column*), and *S. epidermidis* (*right column*) up to 180 min in the presence of AgNP@collagen 1.0 μM and $AgNO_3$ at 1xMIC and 2xMIC. Figure adapted from Alarcon et al. [15]

(non-toxic effects) even at dilutions 2X minimal inhibitory concentration (MIC) when tested with cultures of human skin fibroblasts and epidermal keratinocytes. The collagen-stabilized AgNP were bactericidal against *B. megaterium* and *E. coli* but only bacteriostatic against *S. epidermidis,* see Fig. 1.

In conclusion, although the action mechanism is not completely clear, silver can cause toxicity in multiple sites that is accompanied with leakage of cellular metabolites. It binds on the bacterial cell wall weakens it and eventually leads to its rupture. It promotes cell disruption via hydroxyl radicals and other reactive oxygen species by poisoning respiratory enzymes and components of the microbial electron transport system. Since in the form of AgNP silver is able to diffuse inside the cell, it reacts with thiol containing proteins inactivating them and preferentially binds to DNA to inhibit replication with the concomitant bacteria toxicity.

3 Silver Nanoparticles in Microbiology

As mentioned in Chap. "Biomedical Uses of Silver Nanoparticles: From Roman Wine Cups to Biomedical Devices", the interaction between AgNP and living organisms critically depends on the surface composition of the nanoparticle and the nature of the organism. Thus, the interactions observed for Gram (+) bacteria cannot be directly extrapolated to the observed for Gram (−) due to intrinsic differences in the bacteria outer membrane, see Fig. 2. However, in the following we will revise the main mechanisms for the action of AgNP in bacteria.

Though seemingly trivial, it is worth noting that the exact mode of killing by most antibiotics is as yet unclear despite the long-existent knowledge of their primary targets. It was only recently that the mode of action of the oldest known antibiotic classes, beta lactams, was clearly elucidated [17]. Thus, it is not surprising that the antimicrobial mechanism for AgNP remains not fully understood. However, some lessons can be learned from examining the antimicrobial activity of ionic silver. There are three suggested main mechanisms for the biocidal activities of ionic silver: (i) catalytic oxidation with nascent oxygen to form reactive oxygen species; (ii) complex formation with thiol containing proteins critical in the chain respiratory system [11, 18]; and (iii) binding of silver ions to DNA preventing DNA replication leading to cell cycle arresting and/or death [18–20]. Note that ionic silver is also able to bind RNA of small nuclear antigens [21]. However, it is possible that mechanisms i and ii are interconnected and together assist to iii as revealed in a toxicity study with silver sulfadiazine on Gram negative *P. aeruginosa* that revealed that the silver is distributed within DNA ($\approx$12 %), RNA ($\approx$3 %), lipidic fraction ($<$0.5 %), and the rest within the proteins-polysaccharides fraction [22]. Later studies have however questioned the DNA binding, as the expected silver bound to phosphate groups were not observed in spectroscopic analysis [23]. It may however be the case that the attached bases and not the phosphate backbone provide a site of binding for Ag(I). Interaction of ionic silver with metallothioneins, which are believed to be responsible for silver detoxification in liver, has been studied quite extensively [24].

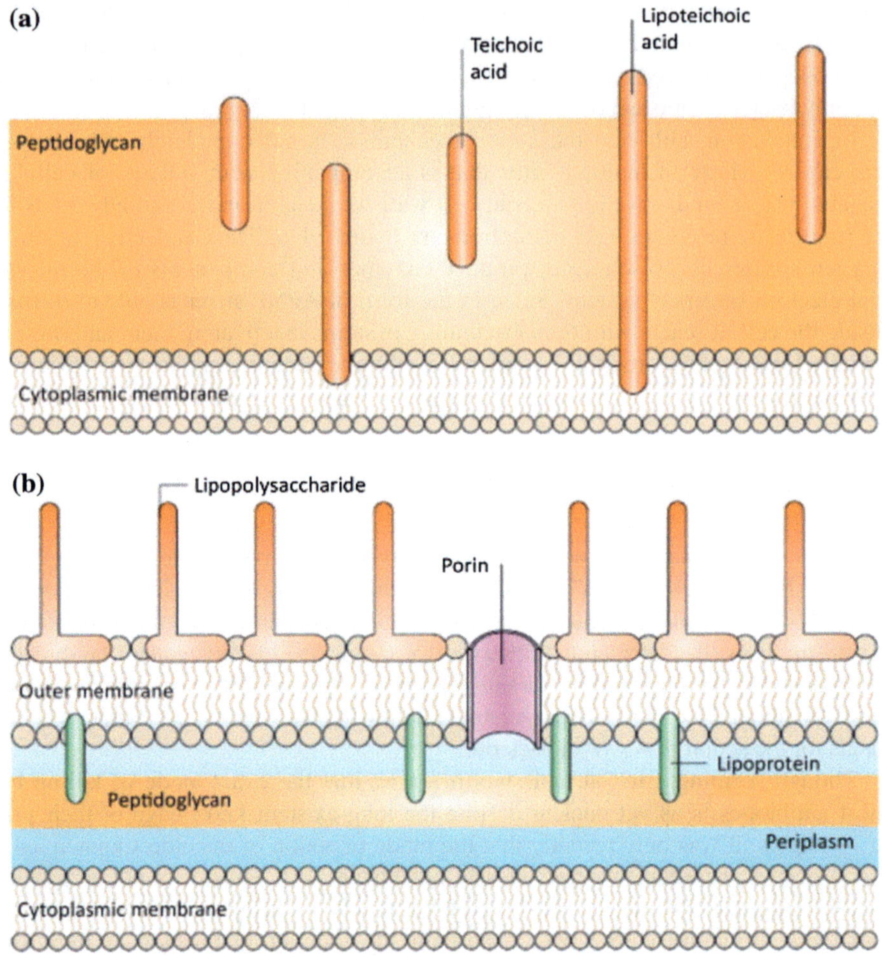

Fig. 2 Schematic representation for bacterial cell structure of: **a** Gram-positive bacterial cell wall that is composed of a thick and multilayered peptidoglycan case. Teichoic acids are connected to and embedded within glycan membrane; **b** Structure for a Gram-negative bacterial cell wall where the outer membrane is linked through lipoproteins to a thin peptidoglycan layer. For these bacteria peptidoglycan layer is placed within the periplasmic space and the outer membrane. Note that the outer membrane contains porins and lipopolysaccharide molecules. Reproduced with permission from Hajipour et al. [16]

Additional damage induced by ionic silver was reported in *Vibrio cholera,* another Gram-negative bacterium. The damage caused is massive proton leakage that results in eventual disruption of membrane potential at sub-micromolar concentrations of ionic silver, resulting ultimately in the death of the microbe [25]. Notably, the study was based on an earlier observation of effects of Ag(I) on ATP production in the (bacterial-derived) mitochondria [26].

The properties of metallic nanoparticles are believed to either differ markedly from the derived metal's ionic state, or on the other extreme, to be exceedingly similar. This dichotomous oversimplification could not be further from the truth, which lies somewhere in between. What has been unclear is whether or not the noted antimicrobial properties are due to the monoionic silver (Ag(I)) or atomic silver Ag(0) leaching out of the nanoparticles. Additionally, were there a distinct mode of action by the AgNP, as compared to $AgNO_3$, the mechanism would need to be elucidated. Some elements of their antimicrobial activity are reminiscent of Ag(I) but these are far from clear. Although some studies have employed transmission electron microscopy (TEM) as a tool to discern AgNP accumulation within the cell, it must be noted that *in-situ* reduction of Ag(I) to Ag(0) can also occur, see Ref. [18]. Sondi and Salopek-Sondi described a study where using TEM, they determined the nature of the interaction between *E. coli* and AgNP of average diameter 12 nm [14]. Their study provided high-resolution images of accumulation of NPs in the cell membranes and identified the formation of what they term 'pits' in the membrane. Going one step further in 2010, Li et al. evaluated the antimicrobial mechanism and activity of silver nanoparticles in *E. coli*, analyzing the growth, permeability and ultrastructure of bacterial cells after treatment with 5 nm AgNP [27]. Their study provided evidence of membrane-disruption and changed permeability as cause of inhibition and/or death in *E. coli*. The extension of these studies could serve to further refine our understanding of the effects specific to AgNP accumulation on specific targets in bacterial cells.

Following the initial TEM studies [14], and confirming a role for membrane disruption, Lok et al., carried out a comprehensive study exploring the antimicrobial action of silver nanoparticles on *E. coli* [28]. Comparing the protein profile obtained from exposure to AgNP with that obtained from exposure ionic silver ($AgNO_3$), they found that *E. coli* cells treated with AgNP present an accumulation of envelope protein precursors that leads to a dissipation of the proton motive force. Note that the synthesized spherical AgNP significantly more effective than Ag(I), being active at nanomolar versus micromolar concentrations (0.4 nM for silver nanoparticles and 6 μM for $AgNO_3$). This was followed soon by another study from the same group that showed the requirement for partially oxidized AgNP with a chemiadsorbed layer of ionic silver adsorbed onto the surface is required for observing antimicrobial activity in *E. coli* [29, 30]. Kim et al., produced spherical silver nanoparticles by borohydride reduction and investigated their antimicrobial properties on yeast, *E. coli* and *S. aureus*. Low concentration of the nanoparticles were effective against *E. coli* and yeast (3.3 and 6.6 nM respectively), on the contrary the growth inhibitory effect on *S. aureus* was only mild [31]. The authors claimed the pivotal role of ROS in the antimicrobial mechanism for AgNP.

Other noteworthy antimicrobial studies carried out with AgNP are those relevant to microbial communities of economic and environmental concern; Choi et al. for instance, evaluated the inhibitory effects of silver nanoparticles, silver ions and silver chloride colloids on mixed nitrifying bacterial communities. Their results showed that AgNP (with average diameters of 14 nm at 1 mg/L) inhibit nitrifying bacteria bacterial growth of *E. Coli* up to 86 %. Silver colloids and silver ions at the same silver content inhibited respiration by 46 and 42 % respectively [32]. The implication of this is that at lower concentrations, AgNP can inhibit collective respiration of the microbes within these communities with almost the double of efficiency. However, no signs of membrane damage in the bacterial wall were observed for any of the silver species in this study.

Ruparelia et al. investigated the antimicrobial properties of silver and copper nanoparticles on four strains of *E. coli*, three strains of *S. aureus* and *B. subtilis* [33]. They performed both solid and liquid phase trials with silver and copper nanoparticles of sizes 3 and 9 nm respectively. The susceptibility in solution varied among the tested microorganisms with *E. coli* and *S. aureus* being more affected by silver nanoparticles in contrast with *B. subtilis* that showed highest sensitivity to copper nanoparticles. Similar results were observed for solid-state diffusion assays. Navarro et al. assessed the short-term toxicity of silver nanoparticles and Ag(I) to photosynthesis of *Chlamydomonas reinhardtii* [34]. They used a polydisperse suspension of AgNP most (centered $\approx$ 25 nm) and to assess the effect on photosynthesis. The toxicity of silver nanoparticles was significantly higher than that of $AgNO_3$ and/or the available Ag(I) present in the solution. The authors concluded that processes involving the post-decomposition of AgNP into Ag(I) mediated by H_2O_2 produced by the algae play a critical role in the elevated toxicity displayed by AgNP.

3.1 Relevance of Size and Shape of Silver Nanoparticles

As discussed in Chap. "Synthetic Routes for the Preparation of Silver Nanoparticles: A Mechanistic Perspective" the wide range of methods that have been used for the synthesis of silver nanoparticles, have given rise to variable particle morphologies including spheres, cubes, rods, wires and multifacets. By definition nanoparticles are smaller than 100 nm and in the case of spherical silver they can contain from 1,000 up to 58,000 atoms, see Table 1 in Chap. "Biomedical Uses of Silver Nanoparticles: From Roman Wine Cups to Biomedical Devices". Silver nanoparticles of different shapes and sizes show unique interactions with bacteria and viruses [30, 35]. As AgNP of smaller size have higher surface area to volume ratios, the relative rates of silver release may be higher for the smaller sized particles, in line with earlier observations. The initial experimental observations were of a diverse range of sizes of AgNP and an early report concluded that only particles of 1–10 nm in diameter exhibited significant antibacterial properties [35]. However to this we should add the possibility that AgNP with sizes can freely permeate inside the cell membrane. Baker and colleagues also showed that AgNP antimicrobial properties were directly related to the total surface area of the nanoparticles [36]. However, the main limitations of those works to clearly discern between pure Ag(I) release and actual role of the material size lie in the difficulty of matching the same number of AgNP per tested volume of cell suspension. Something we call the 'magic' number, which corresponds to the exact number of AgNP per bacteria required to induce cell death.

The antimicrobial efficacy of silver nanoparticles has been confirmed to be shape dependent in studies that utilized differentially shaped nanoparticles and measured the inhibition of bacterial growth suggested that the most effective geometry is truncated triangular silver nanoparticles [35, 37]. In this study it was shown that the triangular geometry needed silver content >1 μg to exert bactericidal properties, while spherical and rod shaped nanopartciles, needed 12.5 and 50–100 μg respectively. These findings were explained in terms of the high-atom-density facets ({111}) that are found in triangular plates when compared to spheres and rods ({100}). Similar conclusions were exposed by Morones and collaborators [35].

In addition, AgNP manufactured using different synthetic techniques and for different purposes may vary in their physicochemical properties, which can lead to significant differences in their biological activity (see Chap. "Biomedical Uses of Silver Nanoparticles: From Roman Wine Cups to Biomedical Devices"). If we also take into consideration possible coatings, stabilizers and/or other hybridized materials it is clear that these additions may lead to modified cellular uptake and altered interactions with biological macromolecules. Consequently, not all AgNP should be considered the same and it is important to understand that adverse reactions that would not be seen in other silver species can arise. Taken together, it appears that the killing efficacy is directly dependent on the rate and location where Ag(I) is being released from the nanoparticles. This release is in turn dependent on the shape and size of the nanoparticle as well as the physicochemical nature of the surroundings.

3.2 Synergistic Activity with Antibiotics

Although there are limited number of studies exploring combinations of silver and antimicrobials, the topic has recently come into more focus. This upsurge in exploring alternative approaches to infection treatment has also raised the possibility of using AgNP with improved pharmacokinetics and even pharmacodynamics. Li et al., synthesized silver nanoparticles by reducing $AgNO_3$ aqueous solution with ascorbic acid aqueous solution with an average size of 20 nm [38]. Exposing *E. coli* to these nanoparticles to a combination with the beta lactam antibiotic amoxicillin, increased the bactericidal efficiency when compared to AgNP or amoxicillin alone. In another study by Shahverdi et al., the combinations of AgNP with five antibiotics (penicillin G, amoxicillin, erythromycin, clindamycin and vancomycin) were tested against both model microbes *E. coli* and *S. aureus* [39]. AgNP were produced by aqueous Ag(I) reduction from the supernatant of a *Klebsiella pneumoniae* culture and their sizes spanned between 5–30 nm. In general, enhanced killing of the bacteria was recorded, when a combinational treatment was employed (AgNP + Antibiotic). Notably, the most prominent effects were observed for vancomycin, amoxicillin and penicillin G against *S. aureus*.

4 Resistance to Silver Compounds

Antibiotic resistance is a major concern worldwide. Bacteria have developed resistance to antibiotics by having evolved various mechanisms including physical removing the antibiotics from cell through efflux pumps, modification of the target site of the antibiotics (e.g. bacterial ribosomal rRNA and proteins [40]) and inactivation of antibiotic through enzymes, alteration to metabolic pathway [41].

Since the emergence of resistant bacteria, the anti-microbial activities of silver have been re-investigated. This time however, AgNPs have been the centre of the renewed interest. The enthusiasm over silver as an alternative to conventional antibiotics comes from early tests showing that AgNPs within the size range of 10–100 nm have strong bactericidal potential against both Gram-positive and Gram-negative bacteria that have reported MDR (Morones et al. [35]). These include strains of MDR Pseudomonas aeruginosa, ampicillin-resistant *E. coli*, erythromycin-resistant *Streptococcus pyogenes*, methicillin-resistant *S. aureus* (MRSA) and vanco-mycin-resistant *S. aureus* (VRSA).

4.1 Silver Ion Resistant Bacteria

In 1975, a strain of *Salmonella typhimurium* was isolated from three burn patients who had been receiving topical treatment with 0.5 % $AgNO_3$ solution [42]. This strain of *S. typhimurium* was found to be resistant to silver nitrate, mercuric chloride, ampicillin, chloramphenicol, tetracycline, streptomycin, and sulphonamides. In this case, other strains isolated showed that they were resistant to the antibiotics but not silver nitrate, suggesting that the resistant *S. typhimurium* could have originated as a a strain that was resistant to the antibiotics, ampicillin and chloramphenicol, but further selected by topical use of $AgNO_3$ solutions on the burned surfaces. The resistance in this case was later attributed to a set of gene Ag(I) resistant genes, *sil,* from a 180 kb plasmid, pMG101, that belongs to the IncH incompatibility class [43]. The region of pMG101 that codes for the inducible silver resistance has now been sequenced and shown to contain seven genes and two open reading frames of unknown function [44]. As described by Simon Silver [44], silver resistance machinery is sort of unique in his class as this is formed by three different resistance mechanisms; a periplasmic multi-metal-binding protein, a chemiosmotic efflux pump and an ATPase efflux pump, all those encoded in a single toxic metal cation resistance gene cluster.

Silver resistant *P. aeruginosa*, has also been isolated in burn patients and the resistance has also been tracked by to a plasmid source [45]. Note that the strains that were also resistant to gentamicin also showed transient resistance to silver, which was lost upon repeated subculture,—while they retained their gentamicin resistance. In addition, it would appear that transfer of plasmids conferring resistance can occur between different bacterial strains by conjugation, e.g. from

Acinetobacter baumannii to *E. coli* [46]. While these cases of bacterial resistance to silver ions and others continue to emerge, they all point to a plasmid-mediated mechanism and from 1975 to 2007, there have been fewer than 20 cases reported [47]. It would appear that products such as dressing that release sub-lethal levels of silver ions, however, may lead to resistance in the bacteria they are intended to block. Overall, however, the incidence of silver resistance remains low compared to that of antibiotic resistance.

4.2 Silver Nanoparticles and Bacterial Resistance

Reports on the bactericidal effects of AgNP on MDR continue to grow in number. AgNP are also currently indiscriminately used in consumer products ranging from odour-resistant socks to dishwashers washing machines, to medical applications as anti-microbials. Despite this, there is paucity in the resistance literature on AgNPs. There is a belief that resistance against AgNP will not evolve as their mechanism of attack against bacteria is through destruction of the cells, circumventing their ability to mutate. Zhang et al. studied the effect of long-term exposure of bacteria to AgNP in a membrane bioreactor [48]. They found that in the bacterial stains studies, there was an overall increase in expression of the silver resistance gene, *silE*, within the population by up to 50-fold at 41 days after exposure to AgNP. However, continued, long-term exposure resulted in a decreased in *silE* expression. However, the lack of reports of resistance to AgNPs may only reflect the enthusiasm for AgNP and not the fact that they do not engender resistance as shown by Zhang et al. even though in this case *silE* expression appeared to have a biphasic response [48].

5 Silver Nanoparticles and Biofilms

Traditionally, microbial research has focused on the study of planktonic cultures, where cells grow often as individuals suspended in a liquid phase. Bacteria can however, also exist in mono-, or multispecies communities of cells often on the interfaces between one phase and another; liquid:solid, liquid:air and solid:air. Several types of these are exceedingly important in clinical settings, particularly venous and urinary catheters. Additionally, they are found to compound the free passage and integrity of waterways, storage reservoirs and play an economically significant role in biofouling of these.

Bacteria can form biofilms on both abiotic and biotic surfaces [49, 50], see Fig. 3 top for a schematic representation. Abiotic surfaces include anything that can be implanted in the body such as urinary or venous catheters, artificial hearts and pacemakers. Biotic surfaces are also susceptible to bacterial colonization and growth. Many interphases exist between liquid, solid and air in the different

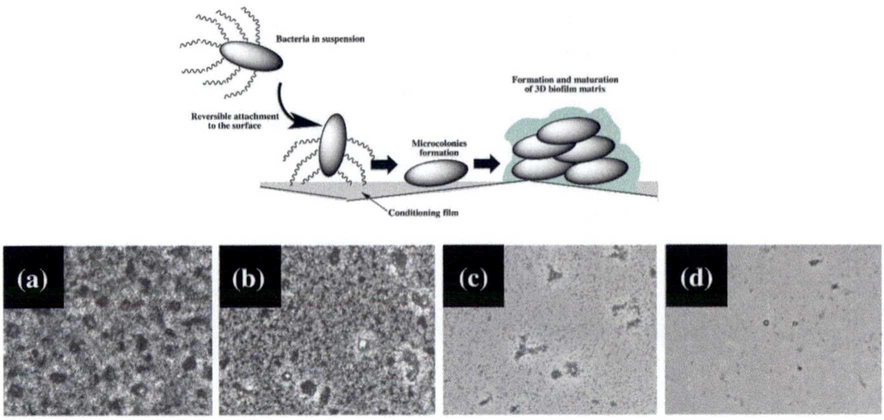

Fig. 3 *Top* Schematic representation for biofilm formation from bacteria culture adapted from Römling and Balsabore [62]. *Bottom* Bright field images showing (*A*) *P. aeruginosa* biofilm that had formed on the surface of 6 well plates after 16 h incubation at 37 °C without any antimicrobials (*control*); and (*B*) Biofilm formed in the presence of 0.156 μM LL37 peptide and in the presence of LL37 coated AgNP (LL37@AgNP) at either 2X and 4X MIC adapted from Vigoni et al. [61]

compartments of the mammalian body. Some of the best-studied examples of infections where biofilms are highly relevant are that of the cystic fibrosis (CF) lung, in recurrent urinary tract infections and chronic diabetic ulcers. Cystic fibrosis afflicted lungs are characterized by the retarded movement of mucus in the lungs. This leads to an increased residence time for inhaled, and otherwise translocated bacteria in the organ and this in its own turn increases the probability of bacterial retention and infection. Biofilms have been shown to play a significant role in the CF-related infections of the airways [51].

It has been estimated that up to 70 % of human bacterial infections are biofilm-related. Once established, biofilms are difficult to eradicate due not only to an increased density of cells and quorum sensing (concerted gene regulation in response to threshold density sensing) related effects inherent in the biofilm, but also reduced permeability of certain antibiotics developing resistance to antibiotics [52, 53]. Due to this much research has focused on trying to identify ways to prevent biofilm formation. There are several promising approaches but one that we will highlight is the use of AgNP.

Based on the results from several studies, it has been demonstrated that AgNP can impact biofilm formation in three ways. In assays where biofilms are formed first and then exposed to AgNP, the AgNP can reduce the numbers of viable cells in the biofilms [54, 55]. Other studies have assessed the ability of AgNP to prevent biofilm formation. When AgNP are added to a bacterial culture at the time of inoculation for a biofilm assay, biofilm formation is inhibited [51, 54, 56–58]. Finally, when AgNP are impregnated into different materials and the amount of biofilm formation by various medically-relevant bacteria is measured, biofilm formation is prevented [59, 60]. In Vigoni et al., AgNP were coated with an anti-microbial, a cathelicidin known as LL-37 [61]. LL-37 has pleiotropic effects, and in this study acted to counteract the cell inhibitory effects of AgNP at higher concentrations. These LL-37 coated AgNP were able to prevent biofilm formation by *Pseudomonas aeruginosa* (*P. aeruginosa*) (Fig. 3). While these studies reported anti-biofilm activities of AgNPs, other studies have reported that silver has no effect, suggesting that the context is important. Furthermore, while the research suggests that the use of AgNP as an anti-biofilm compound is promising, the studies tend to be in vitro. Experimentation must advance into in vivo models to properly assess the therapeutic potential of these molecules.

6 Silver Versus Other Metallic Nanoparticles

As shown above, AgNPs are both bactericidal and bacteriostatic against several bacteria. Other metallic NPs have also been reported to have anti-microbial activity. For example, gold nanoparticles (AuNPs) have documented to be effective anti-microbial agents against *E. coli* and *Salmonella typhi (S. typhi)*, two bacteria that are common water pollutants that are also health hazards [63]. In this study, 5 nm AuNPs were dispersed onto zeolites of clinoptilolite, mordenite and faujasite. Au-faujasite comprising 5 nm Au on the surfaces were reported to eliminate 90–95 % of *E. coli* and *S. typhi* colonies.

A number of studies of AuNPs synthesized using plant products and extracts as reducing agents report anti-microbial activity. The plant products include coriander, *Bischofia javanica, Daucus carota, Solanum lycopersicums, Hibiscus* cannabinus leaf, *Moringa oliefera* flower, lemongrass, *Bacopa monnieri*, *Citrus unshiu* peel and *Ananas comosus* [64].

Of special note is that unless the particles are stabilized, they will oxidise. The oxidized AgO, however, is cytotoxic and no longer effective as an anti-microbial agent [61]. While AgNPs lose their anti-microbial activity when oxidized, metallic oxide nanoparticles (NPs) from zinc (ZnO), copper (CuO) and iron (Fe_2O_3) on the other hand have been shown to possess anti-bacterial activity against both Gram positive and Gram negative bacterial strains [65]. Azam and colleagues showed that the metallic oxide NPs had anti-bacterial activity against Gram-negative bacteria such as *Escherichia coli* (*E. coli*) and *P. aeruginosa*, and Gram-positive *Staphylococcus aureus* (*S. aureus*) and *Bacillus subtilis (B. subtilis*). Patterns of inhibition obtained showed that the anti-bacterial activity observed was positively correlated to an increase in the surface to volume ratio of the NPs, which in this case, was increased as the size of the NPs decreased. Hence efficacy of ZnO > CuO > Fe_2O_3 as demonstrated by the relative zones of inhibition obtained when the different bacterial strains were incubated with the metallic oxide NPs (Fig. 4). In general, the NPs were more effective against Gram-positive strains of bacteria. However, these metallic oxide NPs have also been reported to be cytotoxic, as cautioned by these authors, as reported for silver oxide when AgNPs became oxidized [15].

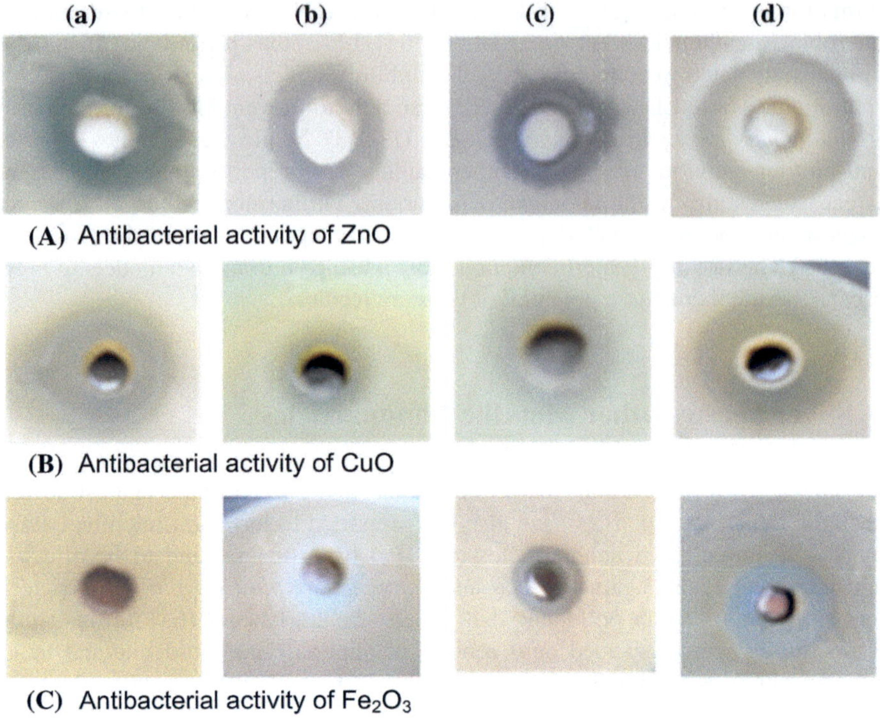

Fig. 4 Zone of inhibition produced by different metal oxide nanoparticles against both Gram-positive and Gram-negative bacterial strains. Antibacterial activity of **a** ZnO; **b** CuO; and **c** Fe_2O_3 of bacterial strains (*a*) *Escherichia coli*, (*b*) *Staphylococcus aureus*, (*c*) *Pseudomonas aeruginosa*, and (*d*) *Bacillus subtilis*. Modified from Azam et al. [65]

7 Pros and Cons of Using Silver Nanoparticles

Not all forms of silver possess antimicrobial properties. Efficacy greatly depends upon the silver species in question, its delivery system, duration of release, and summary bioavailability. Clinically derived pharmacokinetic studies have shown a slower bactericidal activity than common disinfectants such as sodium chlorate and phenol. Also noteworthy from the same study was an observed reduced activity on Gram-positive bacteria [31].

7.1 Side Effects

7.1.1 Ionic Silver

As with any drug, silver ingestion, particularly of mega doses of AgNPs in form of colloidal silver has a number of side effects. Most often, ingestion of the high doses of silver has been found to cause cosmetic abnormalities upon prolonged use. Argyria is an irreversible condition caused by the deposition of silver on skin, a condition that was known since the 1700s. Beyond this discoloration however, no pathologic and other physiological conditions have been correlated with argyria [66–68].

Other clinical indications that have been associated with silver are transient skin discoloration, allergic responses, leucopenia, bone marrow toxicity and renal or hepatic damage due to silver deposition. Silver ions are highly reactive species and readily bind negatively charged proteins, DNA, RNA and other anions, see above. This reactivity complicates the delivery due to the formation of complexes with inflammatory related molecules in the infected site. In topical applications there is evidence that silver ions may form complexes by anions in body fluids (e.g. chloride anions), see Chap. "Biomedical Uses of Silver Nanoparticles: From Roman Wine Cups to Biomedical Devices". The problems with most topical silver antimicrobials lie in the lack of deep tissue penetration, lack of control over release, the limited number of reactive species being released and the slowed down wound healing.

7.1.2 Silver Nanoparticles

Studies suggest that silver is highly toxic to keratinocytes and fibroblasts, with the latter being more sensitive. It is capable of altering gene expression [69]. Moreover, arrested healing in patients treated with silver compounds, due to fibroblast and epithelial cell toxicity, has been reported [70]. Evidence also exists of cytotoxicity of silver nanoparticles towards mammalian germline stem cells [71]. The silver nanoparticles employed in that study drastically reduced mitochondrial function, increased membrane leakage, necrosis and induction of apoptosis. Exposure to silver nanoparticles seems to exert toxicity by decreasing the function of mitochondria.

Silver nanoparticles show very promising results in agar-based screenings but it in liquid medium, even at high concentrations, can only cause a growth delay in *E. coli* [37]. This phenomenon was explained by the coagulation of silver nanoparticles with intracellular substances from the lysed cells, which led to a rapid reduction in available silver in active form [14].

7.2 Advantages

As mentioned previously, the advent of antibiotics was quickly followed by the emergence of antibiotic resistant and later of multi-antibiotic resistant strains that consequently led to a comeback of silver as potential antimicrobial agent. Due to recent technological advances nanosilver production has become increasingly accessible. Nanosilver particles exhibit novel biological, physical and chemical properties that help overcome some of the problems that are associated with earlier metallic silver based treatments. Advances in manufacturing and quality control methods have facilitated the production of AgNP displaying significantly less systemic toxicity.

To add to this, AgNP exhibit a multilevel antibacterial effect on cells and this taken together with the low rates of acquired resistance emergence in many bacterial species, AgNP are particularly promising as antimicrobials. Several studies have highlighted the successful treatment of multidrug resistant species, where last resort antibiotics have failed to contain the infection. The ability to tune the release of the silver reactive species is also being explored in promising studies of long-term bactericidal effects. Further, pioneering work by Wong et al., has shown that materials containing AgNP are able to considerably reduce the inflammation in wounds, allowing for wound healing with minimal tissue scarring [72–74].

8 Anti-Infective Activity on Viruses

AgNP have also been reported to have anti-viral activities against a range of different viral families. The most significant viruses that are tragetted by AgNP include the human immunodeficiency virus (HIV), herpes simplex virus (HSV), and the hepatitis B virus. In many cases, the AgNP were capped or coated with other biomaterials and the effects appear to be synergistic. Table 1 gives a list of viruses inhibited by AgNP together with potential mechanism of action in each case.

Table 1 Anti-viral properties of silver nanoparticles and possible mechanism(s) adapted from Ref. [75]

Virus	Family	Coating	Size (nm)	Mechanism of action
Human immunodeficiency virus type 1 (HIV-1)	*Retroviridae*	PVP coating	1–10	Interaction with gp 120
				Bind with viral envelope glycoprotein
Herpes simplex virus type 1 (HSV-1)	*Herpesviridae*	MES coating	4	Competition for the binding of the virus to the cell
Hepatitis B virus (HBV)	*Hepadnaviridae*	–	10, 50	Interaction with double stranded DNA/binding with viral particles
Monkey pox virus	*Poxviridae*	Simple and polysaccharide coating	10–80	Block of virus-host cell binding and penetration
Tacaribe virus (TCRV)	*Arenaviridae*	Simple and polysaccharide coating	5–10	Inactivation of virus particles before entry
Respiratory syncytial virus	*Paramyxoviridae*	PVP coating	69 ± 3	Interference with viral attachment

9 Concluding Remarks

It is generally accepted that the bactericidal effects of silver nanoparticles are due to ion release. Silver nanoparticles need to address two main functions: to generate a sustained flux of silver ions from a supply of nanoparticles and to be able to actively transport the silver ions to their biological targets on the surface or inside the target cells. An efficient control release mechanism for silver nanoparticles would provide multiple advantages and would become a strong tool to help broaden nanoparticle applications in medicine. Some of the benefits include dose control and limitation in a fashion that the desired bactericidal effects are optimal without enhanced toxicity on host cells, optimization of release kinetics and targeted delivery, control over the nanoparticle lifetime and effectiveness and lastly minimize the additional release of silver ions beyond the therapeutic dose.

Acknowledgments Research in TF Mah's laboratory has been supported by grants from Cystic Fibrosis Canada and the Natural Sciences and Engineering Research Council of Canada (NSERC). EIA thanks the University of Ottawa Heart Institute for the financial and scientific support (UOHI grant#1255). MG acknowledges funding from the Swedish Research Council and AFA Försäkring for research conducted within her laboratory.

References

1. Alexander, J.W.: History of the medical use of silver. Surg. Infect. **10**(3), 289–292 (2009)
2. Roe, A.L., Collosol argentum and its ophthalmic uses. British Med. J. **1**(2820), 104 (1915)
3. Tenover, F.C., Mechanisms of antimicrobial resistance in bacteria. Am. J. Med. **119**(6 Suppl 1), S3–S10; discussion S62–S70 (2006)
4. Rai, M.K., et al.: Silver nanoparticles: the powerful nanoweapon against multidrug-resistant bacteria. J. Appl. Microbiol. **112**(5), 841–852 (2012)
5. Chernousova, S., Epple, M.: Silver as antibacterial agent: ion, nanoparticle, and metal. Angew. Chem. **52**(6), 1636–1653 (2012)
6. Hardes, J., et al.: Lack of toxicological side-effects in silver-coated megaprostheses in humans. Biomaterials **28**(18), 2869–2875 (2007)
7. Butany, J., et al.: Prosthetic heart valves with silver-coated sewing cuff fabric: early morphological features in two patients. Can. J. Cardiol. **18**(7), 733–738 (2002)
8. Butany, J., et al.: Pathologic analysis of 19 heart valves with silver-coated sewing rings. J. Card. Surg. **21**(6), 530–538 (2006)
9. Chopra, I.: The increasing use of silver-based products as antimicrobial agents: a useful development or a cause for concern? J. Antimicrob. Chem. **59**(4), 587–590 (2007)
10. Fox Jr, C.L.: Silver sulfadiazine–a new topical therapy for pseudomonas in burns. Therapy of pseudomonas infection in burns. Arch. Surg. **96**(2), 184–188 (1968)
11. Liau, S.Y., et al.: Interaction of silver nitrate with readily identifiable groups: relationship to the antibacterial action of silver ions. Lett. Appl. Microbiol. **25**(4), 279–283 (1997)
12. Bragg, P.D., Rainnie, D.J.: The effect of silver ions on the respiratory chain of *Escherichia coli*. Can. J. Microbiol. **20**(6), 883–889 (1974)
13. Schreurs, W.J., Rosenberg, H.: Effect of silver ions on transport and retention of phosphate by *Escherichia coli*. J. Bacteriol. **152**(1), 7–13 (1982)
14. Sondi, I., Salopek-Sondi, B: Silver nanoparticles as antimicrobial agent: a case study on E. coli as a model for Gram-negative bacteria. J. Coll. Interface. Sci **275**(1), 177–182 (2004)

15. Alarcon, E.I., et al.: The biocompatibility and antibacterial properties of collagen-stabilized, photochemically prepared silver nanoparticles. Biomaterials **33**(19), 4947–4956 (2012)
16. Hajipour, M.J., et al.: Antibacterial properties of nanoparticles. Trends Biotech. **30**(10), 499–511 (2012)
17. Cho, H., Uehara, T., Bernhardt, T.G.: Beta-lactam antibiotics induce a lethal malfunctioning of the bacterial cell wall synthesis machinery. Cell **159**(6), 1300–1311 (2014)
18. Feng, Q.L., et al.: A mechanistic study of the antibacterial effect of silver ions on *Escherichia coli* and Staphylococcus aureus. J. Biomed. Mater. Res. **52**(4), 662–668 (2000)
19. Batarseh, K.I.: Anomaly and correlation of killing in the therapeutic properties of silver (I) chelation with glutamic and tartaric acids. J. Antimicrob. Chemother. **54**(2), 546–548 (2004)
20. Feng, Q.L., et al.: A mechanistic study of the antibacterial effect of silver ions on *Escherichia coli* and Staphylococcus aureus. J. Biomed. Mater. Res. **52**(4), 662–668 (2000)
21. McNeilage, L.J., Whittingham, S.: Use of the bio-rad silver stain to identify gel purified RNA components of small nuclear ribonucleoprotein antigens. J. Immunol. Met. **66**(2), 253–260 (1984)
22. Modak, S.M., Fox Jr, C.L.: Binding of silver sulfadiazine to the cellular components of Pseudomonas aeruginosa. Biochem. Pharmacol. **22**(19), 2391–2404 (1973)
23. Bovenkamp, G.L., et al.: X-ray absorption near-edge structure (XANES) spectroscopy study of the interaction of silver ions with Staphylococcus aureus, Listeria monocytogenes, and *Escherichia coli*. Appl. Environ. Microbiol. **79**(20), 6385–6390 (2013)
24. Stillman, M.J., et al.: Spectroscopic studies of copper, silver and gold-metallothioneins. Met.-Based Drugs **1**(5–6), 375–394 (1994)
25. Dibrov, P., et al.: Chemiosmotic mechanism of antimicrobial activity of Ag(+) in Vibrio cholerae. Antimicrob. Agents Chemother. **46**(8), 2668–2670 (2002)
26. Chappell, J.B., Greville, G.D.: Effect of silver ions on mitochondrial adenosine triphosphatase. Nature **174**(4437), 930–931 (1954)
27. Li, W.-R., et al.: Antibacterial activity and mechanism of silver nanoparticles on *Escherichia coli*. Appl. Microbiol. Biotech. **85**(4), 1115–1122 (2010)
28. Lok, C.N., et al.: Proteomic analysis of the mode of antibacterial action of silver nanoparticles. J. Proteome Res. **5**(4), 916–924 (2006)
29. Lok, C.N., et al.: Silver nanoparticles: partial oxidation and antibacterial activities. J. Biol. Inorg. Chem. **12**(4), 527–534 (2007)
30. Lok, C.N., et al.: Silver nanoparticles: partial oxidation and antibacterial activities. J. Biol. Inorg. Chem. **12**(4), 527–534 (2007)
31. Kim, J.S., et al.: Antimicrobial effects of silver nanoparticles. Nanomed. Nanotechnol. Biol. Med. **3**(1), 95–101 (2007)
32. Choi, O., et al.: The inhibitory effects of silver nanoparticles, silver ions, and silver chloride colloids on microbial growth. Water Res. **42**(12), 3066–3074 (2008)
33. Ruparelia, J.P., et al.: Strain specificity in antimicrobial activity of silver and copper nanoparticles. Acta Biomater. **4**(3), 707–716 (2008)
34. Navarro, E., et al.: Toxicity of silver nanoparticles to Chlamydomonas reinhardtii. Environ. Sci. Technol. **42**(23), 8959–8964 (2008)
35. Morones, J.R., et al.: The bactericidal effect of silver nanoparticles. Nanotechnology **16**(10), 2346–2353 (2005)
36. Baker, C., et al.: Synthesis and antibacterial properties of silver nanoparticles. J. Nanosci. Nanotechnol. **5**(2), 244–249 (2005)
37. Pal, S., Tak, Y.K., Song, J.M.: Does the antibacterial activity of silver nanoparticles depend on the shape of the nanoparticle? a study of the Gram-negative bacterium *Escherichia coli*. Appl. Environ. Microbiol. **73**(6), 1712–1720 (2007)
38. Li, P., et al.: Synergistic antibacterial effects of β-lactam antibiotic combined with silver nanoparticles. Nanotechnology **16**(9), 1912 (2005)
39. Shahverdi, A.R., et al.: Synthesis and effect of silver nanoparticles on the antibacterial activity of different antibiotics against Staphylococcus aureus and *Escherichia coli*. Nanomed. Nanotechnol. Biol. Med. **3**(2), 168–171 (2007)

40. Musser, J.M.: Antimicrobial agent resistance in mycobacteria: molecular genetic insights. Clin. Microbiol. Rev. **8**(4), 496–514 (1995)
41. Chopra, I., Roberts, M.: Tetracycline antibiotics: mode of action, applications, molecular biology, and epidemiology of bacterial resistance. Microbiol. Mol. Biol. Rev. **65**(2), 232–260 (2001)
42. McHugh, G.L., et al.: Salmonella typhimurium resistant to silver nitrate, chloramphenicol, and ampicillin. Lancet **1**(7901), 235–240 (1975)
43. Gupta, A., et al.: Diversity of silver resistance genes in IncH incompatibility group plasmids. Microbiology **147**(Pt 12), 3393–3402 (2001)
44. Silver, S.: Bacterial silver resistance: molecular biology and uses and misuses of silver compounds. FEMS Microbiol. Rev. **27**(2–3), 341–353 (2003)
45. Bridges, K., et al.: Gentamicin- and silver-resistant pseudomonas in a burns unit. Br. Med. J. **1**(6161), 446–449 (1979)
46. Deshpande, L., Chopade, B.: Plasmid mediated silver resistance in Acinetobacter baumannii. Biometals **7**(1), 49–56 (1994)
47. Chopra, I.: The increasing use of silver-based products as antimicrobial agents: a useful development or a cause for concern? J. Antimicrob. Chemother. **59**(4), 587–590 (2007)
48. Zhang, C., Liang, Z., Hu, Z.: Bacterial response to a continuous long-term exposure of silver nanoparticles at sub-ppm silver concentrations in a membrane bioreactor activated sludge system. Water Res. **50**, 350–358 (2014)
49. Hall-Stoodley, L., Costerton, J.W., Stoodley, P.: Bacterial biofilms: from the natural environment to infectious diseases. Nat. Rev. Micro. **2**(2), 95–108 (2004)
50. Romling, U., et al.: Microbial biofilm formation: a need to act. J. Intern. Med. **276**(2), 98–110 (2014)
51. Palanisamy, N.K., et al.: Antibiofilm properties of chemically synthesized silver nanoparticles found against Pseudomonas aeruginosa. J. Nanobiotech. **12**, 2 (2014)
52. Mah, T.F.: Biofilm-specific antibiotic resistance. Future Microbiol. **7**(9), 1061–1072 (2012)
53. Hoiby, N., et al.: Antibiotic resistance of bacterial biofilms. Int. J. Antimicrob. Agents **35**(4), 322–332 (2010)
54. Martinez-Gutierrez, F., et al.: Anti-biofilm activity of silver nanoparticles against different microorganisms. Biofouling **29**(6), 651–660 (2013)
55. Park, H.-J., et al.: Biofilm-inactivating activity of silver nanoparticles: a comparison with silver ions. J. Ind. Eng. Chem. **19**(2), 614–619 (2013)
56. Kalishwaralal, K., et al.: Silver nanoparticles impede the biofilm formation by Pseudomonas aeruginosa and Staphylococcus epidermidis. Colloids Surf. B **79**(2), 340–344 (2010)
57. Mohanty, S., et al.: An investigation on the antibacterial, cytotoxic, and antibiofilm efficacy of starch-stabilized silver nanoparticles. Nanomedicine **8**(6), 916–924 (2012)
58. Radzig, M.A., et al.: Antibacterial effects of silver nanoparticles on gram-negative bacteria: influence on the growth and biofilms formation, mechanisms of action. Colloids Surf. B **102**, 300–306 (2013)
59. Babapour, A., et al.: Low-temperature sol-gel-derived nanosilver-embedded silane coating as biofilm inhibitor. Nanotechnology **22**(15), 155602 (2011)
60. Paladini, F., et al.: Efficacy of silver treated catheters for haemodialysis in preventing bacterial adhesion. J. Mater. Sci. Mater. Med. **23**(8), 1983–1990 (2012)
61. Vignoni, M., et al.: LL37 peptide@silver nanoparticles: combining the best of the two worlds for skin infection control. Nanoscale **6**(11), 5718–5725 (2014)
62. Romling, U., Balsalobre, C.: Biofilm infections, their resilience to therapy and innovative treatment strategies. J. Intern. Med. **272**(6), 541–561 (2012)
63. Lima, E., et al.: Gold nanoparticles as efficient antimicrobial agents for *Escherichia coli* and Salmonella typhi. Chem. Cent. J. **7**(1), 11 (2013)
64. Bindhu, M.R., Umadevi, M.: Antibacterial activities of green synthesized gold nanoparticles. Mat. Lett. **120**, 122–125 (2014)
65. Azam, A., et al.: Antimicrobial activity of metal oxide nanoparticles against Gram-positive and Gram-negative bacteria: a comparative study. Int. J. Nanomed. **7**, 6003–6009 (2012)

66. Bouts, B.A.: Images in clinical medicine. Argyria. N. Engl. J. Med. **340**(20), 1554 (1999)
67. Hanada, K., et al.: Silver in sugar particles and systemic argyria. Lancet **351**(9107), 960 (1998)
68. Legat, F.J., et al.: Argyria after short-contact acupuncture. Lancet **352**(9123), 241 (1998)
69. Poon, V.K., Burd, A.: In vitro cytotoxity of silver: implication for clinical wound care. Burns **30**(2), 140–147 (2004)
70. Burd, A., et al.: A comparative study of the cytotoxicity of silver-based dressings in monolayer cell, tissue explant, and animal models. Wound Rep. Reg. **15**(1), 94–104 (2007)
71. Braydich-Stolle, L., et al.: In vitro cytotoxicity of nanoparticles in mammalian germline stem cells. Toxicol. Sci. **88**(2), 412–419 (2005)
72. Liu, X., et al.: Silver nanoparticles mediate differential responses in keratinocytes and fibroblasts during skin wound healing. Chem. Med. Chem. **5**(3), 468–475 (2010)
73. Zhang, S., et al.: Silver nanoparticle-coated suture effectively reduces inflammation and improves mechanical strength at intestinal anastomosis in mice. J. Pedia. Surg. **49**(4), 606–613 (2014)
74. Wong, K.K.Y., et al.: Further evidence of the anti-inflammatory effects of silver nanoparticles. Chem. Med. Chem. **4**(7), 1129–1135 (2009)
75. Khandelwal, N., et al.: Application of silver nanoparticles in viral inhibition: a new hope for antivirals. Digest J. Nanomater. Biostruct. **9**(1), 175–186 (2014)

Erratum to: Biomedical Uses of Silver Nanoparticles: From Roman Wine Cups to Biomedical Devices

Hasitha de Alwis Weerasekera, May Griffith and Emilio I. Alarcon

Erratum to:
Chapter 'Biomedical Uses of Silver Nanoparticles: From Roman Wine Cups to Biomedical Devices' in: E.I. Alarcon et al. (eds.), *Silver Nanoparticle Applications*, Engineering Materials, DOI 10.1007/978-3-319-11262-6_5

The original version of this article was published with some incorrect values. The correct information is given below:

Page 97, line no. 7, "…0.006…" should read as "…0.06…"
Page 97, line no. 10, "…0.6 %…" should read as "…6 %…"

The online version of the original chapter can be found under
DOI 10.1007/978-3-319-11262-6_5

H. de Alwis Weerasekera
Department of Chemistry and Centre for Catalysis Research and Innovation, University of Ottawa, Ottawa, Canada

M. Griffith (✉)
Department of Clinical and Experimental Medicine, Integrative Regenerative Medicine Centre, Linköping University, Linköping, Sweden
e-mail: may.griffith@liu.se

E.I. Alarcon (✉)
Bio-nanomaterials Chemistry and Engineering Laboratory, Division of Cardiac Surgery, University of Ottawa Heart Institute, 40 Ruskin Street Rm H5229, Ottawa K1Y 4W7, Canada
e-mail: ealarcon@ottawaheart.ca

E.I. Alarcon
Centre for Catalysis Research and Innovation, University of Ottawa, Ottawa, Canada

E.I. Alarcon et al. (eds.), *Silver Nanoparticle Applications*, Engineering Materials,
DOI 10.1007/978-3-319-11262-6_7

MIX
Papier aus verantwortungsvollen Quellen
Paper from responsible sources
FSC® C105338

If you have any concerns about our products,
you can contact us on
ProductSafety@springernature.com

In case Publisher is established outside the EU,
the EU authorized representative is:
Springer Nature Customer Service Center GmbH
Europaplatz 3, 69115 Heidelberg, Germany

Printed by Libri Plureos GmbH
in Hamburg, Germany